95 DEL
D1124629

INVERTEBRATE PHOTORECEPTORS

A Comparative Analysis

INVERTEBRATE PHOTORECEPTORS

A Comparative Analysis

JEROME J. WOLKEN

BIOPHYSICAL RESEARCH LABORATORY
CARNEGIE INSTITUTE OF TECHNOLOGY
CARNEGIE-MELLON UNIVERSITY
PITTSBURGH, PENNSYLVANIA

1971

ACADEMIC PRESS ■ New York and London

OPTOMETRY LIBRARY

COPYRIGHT © 1971, BY ACADEMIC PRESS, INC.
ALL RIGHTS RESERVED
NO PART OF THIS BOOK MAY BE REPRODUCED IN ANY FORM,
BY PHOTOSTAT, MICROFILM, RETRIEVAL SYSTEM, OR ANY
OTHER MEANS, WITHOUT WRITTEN PERMISSION FROM
THE PUBLISHERS.

ACADEMIC PRESS, INC.
111 Fifth Avenue, New York, New York 10003

United Kingdom Edition published by
ACADEMIC PRESS, INC. (LONDON) LTD.
Berkeley Square House, London W1X 6BA

LIBRARY OF CONGRESS CATALOG CARD NUMBER: 73-127709

PRINTED IN THE UNITED STATES OF AMERICA

QL 364
W64
OPTOMETRY
LIBRARY

"Don't bite my finger, look where I am pointing"

Embodiments of Mind, W. S. McCulloch
MIT Press, Cambridge, Massachusetts, 1965

This book is dedicated to the memory of my friend
WARREN S. McCULLOCH

3059

CONTENTS

Preface ix
Acknowledgments xi

I. Photobiology

Introduction 1
Radiation 1
The Nature of Light 3
Photoreceptors 4
Photosensitivity and Pigments 5
Experimental Methods 17

II. The Protozoan Photoreceptor: Eyespot and Flagellum

Protozoa 20
Euglena gracilis 23
Photomotion 27
Photoreceptor Pigment 36
The Flagellum and Excitation 41

III. The Compound Eye

Imaging Eyes 47
Compound Eye Structure 50
Insect Compound Eyes 52
Related Discussion 65

IV. Crustacea and Mollusc Eyes

Crustacea 70
Molluscs 86
The Rhabdom Structure 89

V. The Vertebrate Retinal Photoreceptors

The Retinal Rod 93
Vertebrate Visual Pigments 100
Molecular Structure of the Retinal Rod 107

VI. The Invertebrate Eye and Its Visual Pigments

The Visual Pigments 112
Accessory and Screening Pigments 133
Concluding Remarks 137

VII. Summary and Concluding Thoughts

Primitive Photoreceptors 140
Visual Photoreceptors 141
Photoreceptor Molecules 148
Spectral Sensitivity, Pigments, and Color Vision 149
Polarized Light Analysis 151
Photoreceptor Evolution 154

References 156

Appendix 170

Subject Index 173

PREFACE

To obtain information about the photoreceptor systems of plants and animals, I began more than a decade ago to investigate invertebrate animals, their photobehavior, photoreceptor structure, and photopigments. The invertebrates selected, although primarily from those studied in my laboratory, are indicative of the variety of photoreceptors found in the invertebrate phyla. Other animals were included either because their photoreceptor systems were unique or served as models for exploring invertebrate photoreception. The presentation is illustrative rather than exhaustive.

In this comparative study, the structure and pigment chemistry of invertebrate photoreceptors were studied in an effort to discover their function. I have tried to present a coherent story, beginning with the photoreceptor system of protozoa with their eyespot and flagellum, followed by, in phylogenetic order, the compound eye of arthropods (including insects and crustacea), and ending with the refracting eye of molluscs. Analogies have been drawn between these findings for invertebrate photoreceptors and those for the vertebrate visual system.

In addition to the specific reference citations, a Supplemental Readings list has been included to fill in the gaps and omissions. For additional information regarding the phylogenetic position of the invertebrate animals, which are included in the text, the reader is referred to Lord Rothchild's, "A Classification of Living Animals" (Wiley, 1961) and L. H. Hyman's, "The Invertebrates," Volumes I–VI (McGraw-Hill, 1940–1967).

This monograph is not a review of invertebrate photobiology or all invertebrate photoreceptor systems, rather it is a personal account of the photoreceptors I have studied. Hopefully, the reader will find this summary an attempt to extend our understanding of invertebrate photoreceptors to the limits of molecular biology.

JEROME J. WOLKEN

ACKNOWLEDGMENTS

I would like to thank all who have been associated with the Biophysical Research Laboratory and who shared in these studies.

I would like to acknowledge the assistance of A. Jonathan Wolken for his helpful suggestions toward shaping this book in its final form.

My special thanks go to Miss Arlene R. Mann for her skill in getting numerous drafts typed and organized into a readable text. I would like to thank the staff of Academic Press for skillful editorial assistance in seeing this book through from manuscript to publication.

For permission to reproduce Figures 1.8, 1.9, 2.18b, 4.5a, 4.11a, 4.19, 6.1a, 6.4, and 6.20, I would like to thank Dr. K M. Hartman, University of Frieberg; Dr. S. B. Hendricks, U.S.D.A., Beltsville; Dr. G. Tollin, University of Arizona; Dr. T. H Waterman, Yale University; Dr. N. Moray, University of Toronto; Dr. H Zonana, Yale University; Dr. H. Autrum, University of Munich; Paul Brown, Harvard University; and Dr. G. K. Strother, Pennsylvania State University.

I would like to acknowledge my thanks to C. C. Thomas and Co., Springfield, Illinois; Appleton-Century-Crofts, New York; D. Van Nostrand Reinhold, New York; and The Rockefeller University Press Journals, New York for permission to reproduce figures from my previous publications. Material taken or adapted from other sources are acknowledged in the figures and tables, as well as in the references. For permission to use this information, I am most appreciative.

Finally, I am most grateful to the National Aeronautics and Space Administration (NASA) and to the Scaife Family of Pittsburgh for their interest and financial support of this research.

I. PHOTOBIOLOGY

Introduction

All plants and animals, from bacteria to man, show some form of photosensitivity. This photosensitivity is exhibited in behavior as *phototropism* and *phototaxis,* the bending, moving, or swimming to or away from the light stimulus; *photosynthesis,* the conversion of light energy to chemical energy in the synthesis of sugars and starches; *vision,* the conversion of light energy to chemical and electrical energy in the retina of the eye; and *hormonal stimulation,* including growth, sexual cycles, flowering of plants, color changes, and other photobehavioral phenomena.

Before we begin to discuss the photobehavior and photoreceptor systems of invertebrate animals, we have to arrive at a basis by which we hope to understand them. To do so will require knowledge of: the nature of light and its interaction with matter; the plant and animal photoreceptors that are known; and the general kinds and structure of pigment molecules that are identified with these photoprocesses. From this information we can deduce relationships between photoreceptors and their structure and function.

Radiation

The spectrum of electromagnetic radiation extends from gamma rays less than 0.01 Å long, to radio waves several kilometers long (Figure 1.1). However, all photobiological phenomena such as plant and animal phototropism, phototaxis, photosynthesis, and vision take place in the visible part of the spectrum, a very narrow band from about 3900 Å to about

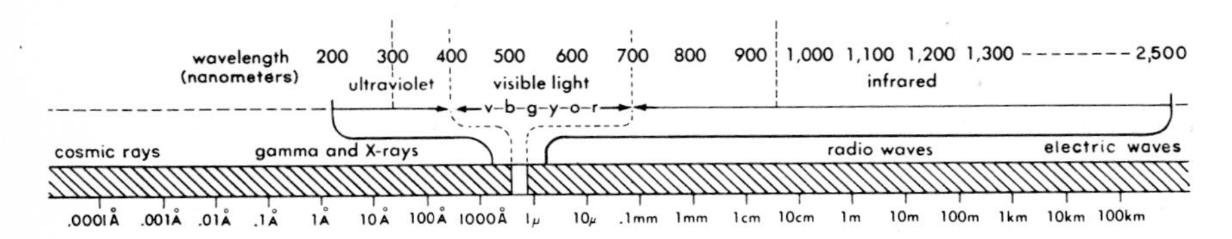

Figure 1.1. The electromagnetic spectrum (10 Å = 1.0 nm = 10^{-3} μ = 10^{-7} cm).

7600 Å (Figure 1.1). The spectrum of solar radiation that reaches the surface of the earth covers only this range, with a maximum around 5000 Å, about which the photobiological phenomena cluster (Figure 1.2).

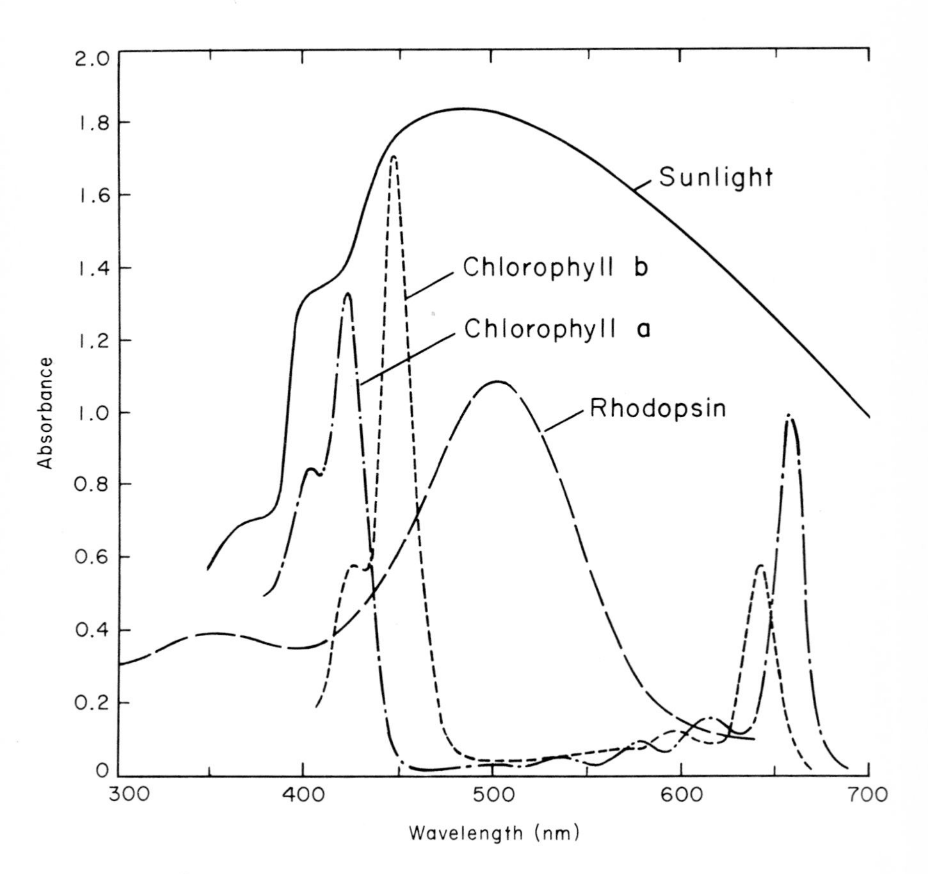

Figure 1.2. The spectrum of sunlight (in relative energy) that strikes the earth's surface compared with the absorption spectra of the photosynthetic pigments chlorophyll a, chlorophyll b, and the visual pigment rhodopsin.

Ultraviolet radiation below 3000 Å is largely absorbed by ozone in the upper atmosphere, but when absorbed by proteins and nucleic acids of living cells, mutagenic and damaging reactions occur. Exactly what part ultraviolet radiation from 3400 to 3900 Å plays in photoprocesses is not clear. Action spectra for the phototropism of lower plants and animals and the visual spectral sensitivity of most insects show a major response peak near 3600 Å. Since action spectra are indicative of the absorption spectrum of the molecules involved, these organisms must possess molecules to absorb this energy.

Radiation beyond 6000 Å in the infrared is important for plant and animal growth, the timing of plant flowering, sexual cycles in animals, and pigment migration. The timing of flowering cycles in plants, called *photoperiodism,* is controlled in the near red part of the spectrum by the shifting of light between 6600 Å and 7300 Å. Bacterial photosynthesis takes place even further in the red, beyond 8000 Å. Infrared radiation beyond 9000 Å is mostly absorbed by atmospheric water vapor and by water that surrounds living cells. Therefore, the limits of radiation effective for photobiology are considered to lie between 3000 and 9500 Å (Clayton, 1965; Wald, 1965).

The Nature of Light

In the interactions between light and matter, electromagnetic radiation sometimes behaves as though it were composed of discrete particles. These particles are called *photons* or *quanta* and represent the packets of energy that comprise any type of electromagnetic radiation. A particular type of radiation is characterized on the basis of either its wavelength or its energy. Max Planck discovered in 1900 the direct relationship between the frequency of electromagnetic energy and the energy of its quanta. Albert Einstein extended Planck's relationship to include light. Accordingly, the energy of a single quantum can be calculated from:

$$E = h\nu$$

where E is the energy of the photon, h is Planck's constant (6.625×10^{-27} erg-second), and ν is the frequency of the electromagnetic radiation. This equation shows that the higher the frequency of the radiation, the greater the energy. Since frequency is inversely proportional to wavelength, the shorter the wavelength, the greater the energy. Thus all light quanta of a given wavelength or corresponding frequency have exactly the same amount of energy. Einstein postulated that all the energy of a single light quantum, or photon, is transferred to a single electron. This one-to-one relationship between a light quantum and a particle of matter is of key importance in photochemistry. The principle that one quantum of light can bring about a direct photochemical change in exactly one molecule of matter is known as Einstein's Law of Photochemistry.

Photoreceptors

Photoreceptors are specialized organelles of cells containing photosensitive pigment systems structured for energy capture and transfer. That is, they convert light energy to chemical or electrical energy in the process of photoexcitation. The photoreceptor structures for photosynthesis are the *chloroplasts;* in photosynthetic bacteria they are called *chromatophores,* and in algae, *plastids.* The photoreceptors for vertebrate vision are the retinal *rods* and *cones* of the eye. In the invertebrates, which include protozoa, coelenterates, flatworms, arthropods, and molluscs, the photoreceptors are *eyespots, photosensory cells, ocelli,* and image-forming *compound eyes.* Crustacea and molluscs, as well as cold-blooded vertebrates, possess *chromatophores* (not to be confused with photosynthetic chromatophores), which are yellow, brown, and red pigmented granules, but if black they are called *melanophores.* Collectively, the chromatophores produce color and shade changes in the skin of the animal. For many of these animals the eye is the receptor organ which, through hormonal action, initiates the expansion and contraction of the chromatophores that bring about these color changes. A variation of this pigment effector system occurs in cases in which the pineal organ (gland) is photosensitive, and in lower vertebrates, e.g., amphibia and lizards, participates in the control of adaptive pigmentation (Wurtman *et al.,* 1968). The pineal also bears a strong morphological resemblance to the vertebrate retinal rods and cones (Eakin, 1968, see Figure 5.5).

Eyes are not the sole means of photoreception, for the general body surface of many eyeless and blinded animals is known to be remarkably light-sensitive. This diffuse photosensitivity is defined as the *dermal light sense.* Many invertebrate animals have no recognizable eyes, but have a diffuse photosensitivity over the whole or part of their body (Millot, 1968). For example, in the clam, *Mya,* a sudden change of light intensity results in retraction of the siphons. In the earthworm, the surface of the anterior segments is light-sensitive. These animals possess large numbers of *photosensory cells* located beneath their skin. In almost every phylum, including protozoa and coelenterates, some kind of "eye" has developed that enables the animal to detect the direction and intensity of light. There is evidence to show that deeper tissue cells can also be photosensitive, as, for example, in certain marine animals which exhibit swimming responses if their nerve or ganglion cells are exposed to light (Kennedy, 1963).

What is important to note here is not so much the form of the photoreceptor – that is, whether it is an eyespot, photosensory cell, compound eye, or retinal rod—but rather the unique sensitivity to light each photoreceptor exhibits. There are indications that the process of photoexcitation exhibits a pattern of activity which is fundamentally the same over a wide range of photobiological phenomena, and that the photoreception process may be generally formulated as follows:

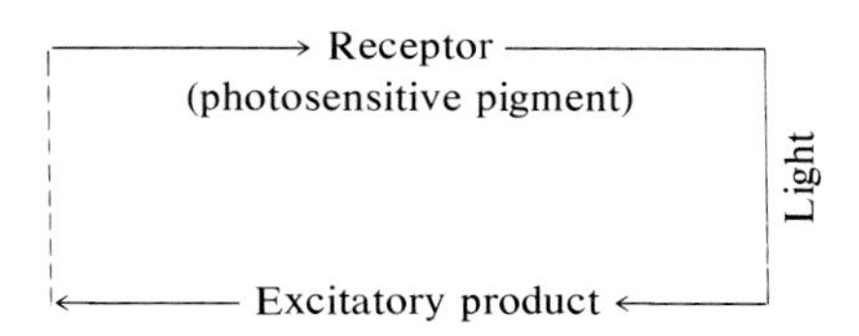

Photosensitivity and Pigments

All photophenomena depend on the ability of a system to absorb light energy. To accomplish this, specific molecules, *pigments*—or an aggregate of interacting molecules called a pigment system—are required to absorb light of the necessary wavelengths. The two most abundant pigments found in nature are the green chlorophylls and the yellow-orange-red carotenoids.

Pigments

The photosynthetic pigments are the chlorophylls. Chlorophyll is a cyclic tetrapyrrole which has the empirical formula $C_{55}H_{72}O_5N_4Mg$, and its "greenness" comes from the magnesium atom at the nucleus of the molecule. Its molecular structure has been described as tadpole-like, with a porphyrin "head" and a phytol "tail" (Figure 1.3). One of the biosynthetic schemes for the synthesis of chlorophyll and heme pigments developed by Granick (1948, 1950, 1958) is presented in Figure 1.4.

Of the various chlorophyll isomers, chlorophyll a and chlorophyll b are found in all higher plants. Chlorophyll a differs from chlorophyll b by possessing a methyl group at the third carbon (Figures 1.3 and 1.4), whereas in chlorophyll b a formyl (—CHO) group occupies this position; chlorophyll b is therefore an aldehyde of chlorophyll a. Chlorophyll a and b differ in absorption spectra (Figure 1.2) as well as in solubility. For example, chlorophyll b is more soluble in methyl alcohol whereas chlorophyll a is more soluble in petroleum ether. These differences make it possible to separate the two chlorophylls. Chlorophyll a is present in all

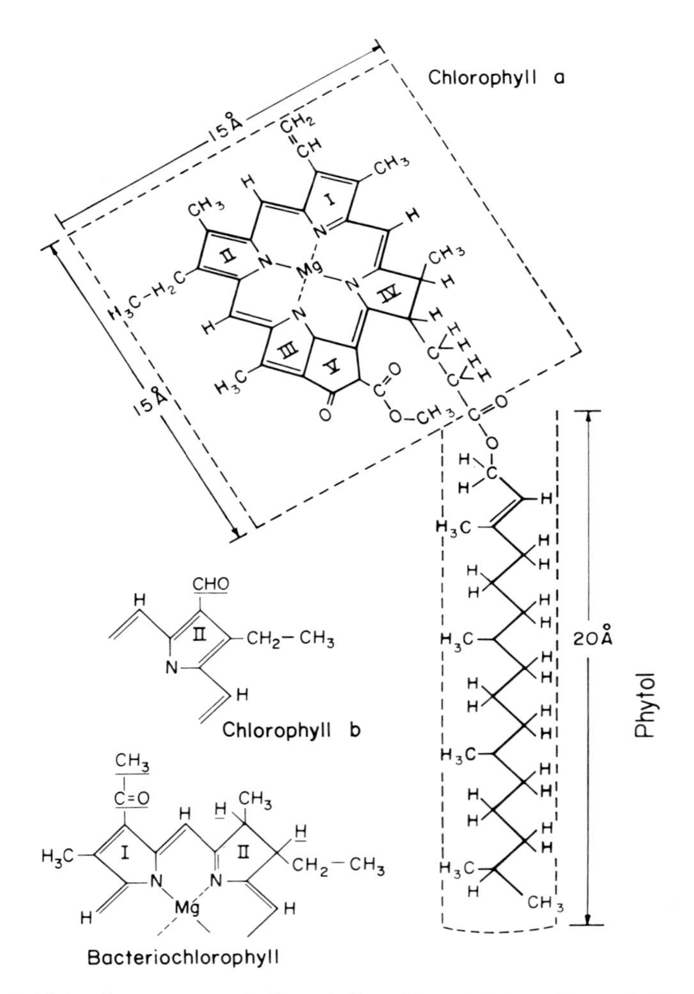

Figure 1.3. Molecular structure of chlorophyll a, chlorophyll b, and bacteriochlorophyll.

green plants except photosynthetic bacteria, while chlorophyll b is found in all higher plants, ferns, mosses, green algae, and euglenoids.

The other chlorophyll isomers are chlorophyll c, d, and e. Chlorophyll c, which lacks a phytol group, is soluble in aqueous alcohols and is found in diatoms, brown algae, dinoflagellates, crytomonads, and chrysomonads. Chlorophyll d is believed to be an oxidation product of chlorophyll a in which the vinyl group at position 2 is oxidized to a formyl group. It is found, together with chlorophyll a, in most red algae. Chlorophyll e,

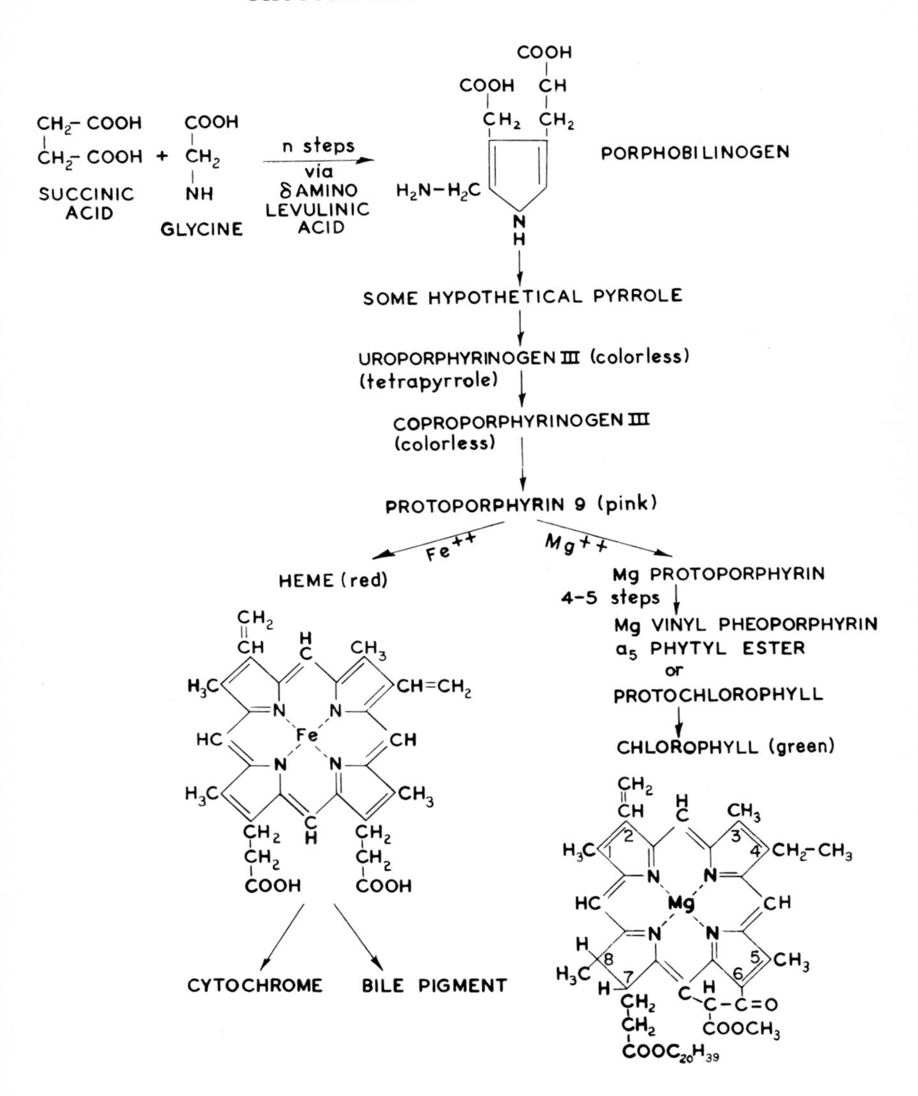

Figure 1.4. Scheme for the biosynthesis of porphyrins, hemes, and chlorophyll (after Granick, 1948, 1950, 1958).

together with chlorophyll a, is present in small amounts in yellow-green algae.

Pheophytin a and b are degradation products of chlorophyll a and b from which the magnesium has been removed. Pheophytin is obtained by acidifying chlorophyll extracts, and its formation can be observed by an accompanying color shift from green to yellow.

Bacteriochlorophyll is found free of other chlorophylls in photosynthetic purple bacteria. It differs from chlorophyll a in that the vinyl group at position 2 is replaced by an acetyl group, and in that it contains two extra hydrogen atoms at positions 3 and 4 (Figure 1.3).

Seeds and etiolated plants (seedlings sprouted in darkness) contain no chlorophyll. However, upon exposure to light, they will turn green: the substance responsible for this reaction is protochlorophyll, the chlorophyll precursor (Figure 1.4). Protochlorophyll differs from chlorophyll in that it lacks two hydrogen atoms at positions 7 and 8 in the porphyrin part of the molecule. Thus it is an oxidation product of chlorophyll a (Figures 1.3 and 1.4).

The carotenoids are yellow, orange to red, fat-soluble pigments that are widely distributed in plant and animal cells. Generically they are named for their most familiar substance, carotene, and are divided into two main groups: *carotenes* (hydrocarbons), the most abundant of which is all-*trans*-β-carotene, $C_{40}H_{56}$ (Figure 1.5); and the *xanthophylls* (oxygen-containing derivatives). One of the more common xanthophylls is lutein, $C_{40}H_{56}(OH)_2$, or luteol (Figure 1.6). From the structure elucidated by Karrer (Karrer and Jucker, 1950), carotenoids can be considered to be built from isoprene units. The linear portion of the molecule consists of four radicals of isoprene (2-methyl-1,3-butadiene) residues. The isoprene units are linked so that the two methyl groups nearest the center of the molecule are in positions 1 and 6, while all other lateral methyl groups are in positions 1 and 5. This may be seen in the structure of several carotenoids presented in Figure 1.6.

The carotenoid molecule, then, is made up of a chromophoric system of alternating single and double interatomic linkages—called conjugated double bonds—between the carbon atoms of a long chain. The carotenoids possess 40 carbon atoms in each molecule.

The large number of these conjugated double bonds offers the possibility of either *cis* or *trans* geometric configurations. Zechmeister (1944, 1962) calculated that there are twenty possible geometric isomers of β-carotene, of which six *cis* isomers have been discovered in nature. The spectral characteristics, and therefore the color of the carotenoids, are largely determined by the number of conjugated double bonds in the molecule (Figures 1.5–1.7).

The biosynthesis of carotenoids is generally associated with the 20-carbon, aliphatic alcohol phytol (Figure 1.6), which is the colorless

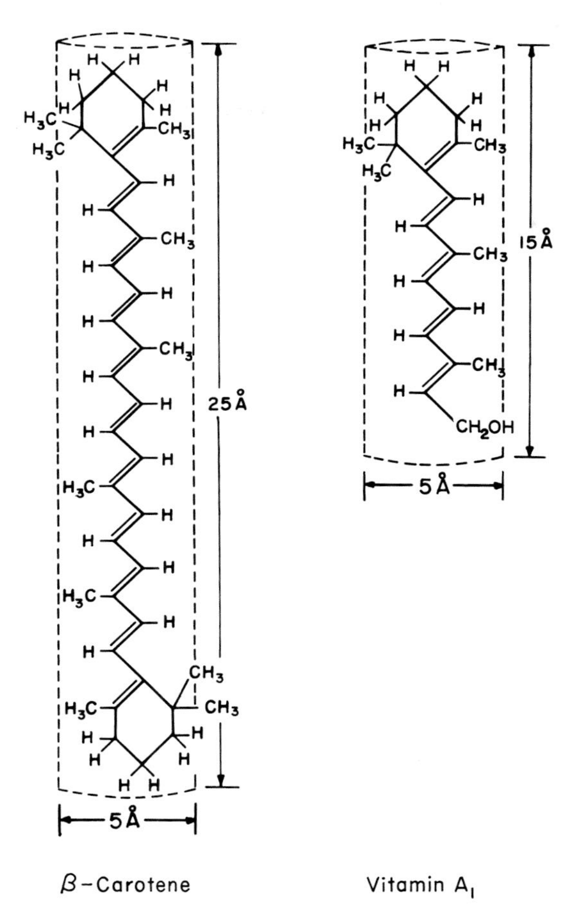

Figure 1.5. Structure of β-carotene and vitamin A_1.

moiety of the ester-comprising chlorophyll (Figure 1.3). The striking resemblance between the carotenoid skeleton and phytol is also true for the details of spatial configuration.

Carotenoids are easily and abundantly synthesized by plants. However, an important change occurred in the development of animals. Animals became dependent on the ingestion of plants for their source of carotenoids. Note that it is not the ingested plant carotenoids but rather the degraded derivative vitamin A that is necessary for all animal life. Thus the carotenoids can be seen to play a central role in the biochemical evolution

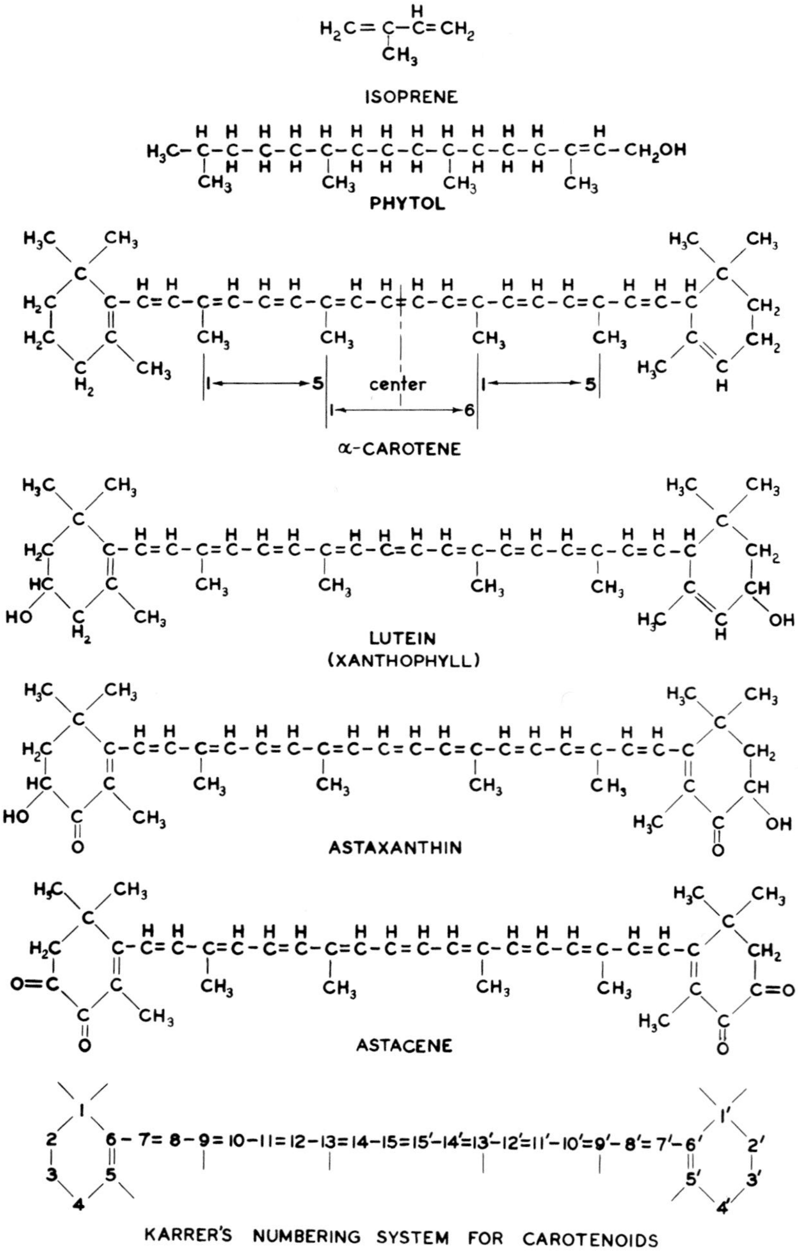

ISOPRENE

PHYTOL

center

α-CAROTENE

LUTEIN
(XANTHOPHYLL)

ASTAXANTHIN

ASTACENE

KARRER'S NUMBERING SYSTEM FOR CAROTENOIDS

Figure 1.6. Molecular structures of carotenoids.

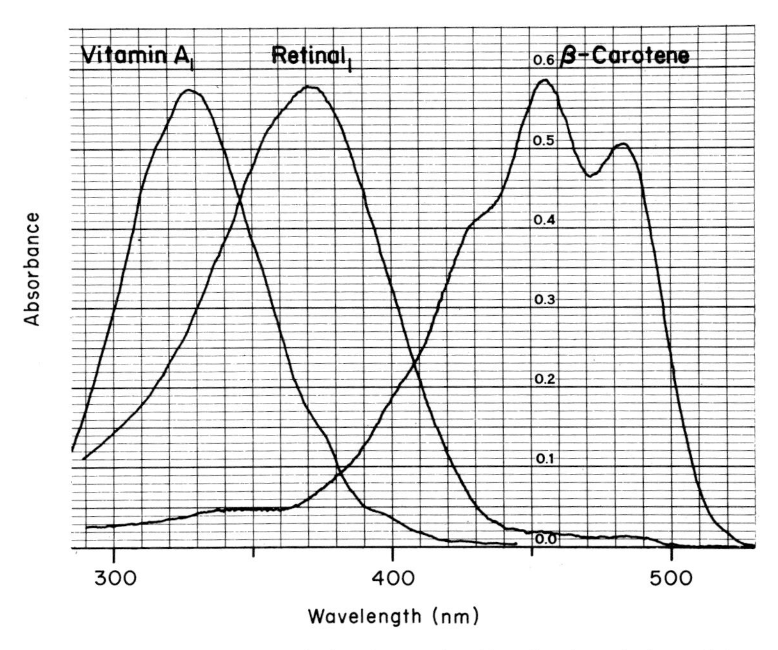

Figure 1.7. Absorption spectra of vitamin A_1 (in chloroform), retinal$_1$, and β-carotene (in benzene).

from the *plant* C_{40} (β-carotene) → *animal* C_{20} (vitamin A). The importance of this evolution to our understanding of the visual pigment chemistry will be elaborated in more detail in Chapter V, with regard to the visual pigment of vertebrate eyes, and in Chapter VI, with regard to the invertebrate visual pigments.

A more detailed discussion of the biosynthesis and chemical structure of the carotenoids and their distribution throughout the plant and animal kingdom can be found in: Fox (1953); Goodwin (1952, 1965); Karrer and Jucker (1950); Strain (1951); and Zechmeister (1944, 1962).

As previously indicated, photoperiodism—the response to variations in the wavelength of light and the length of day and night—controls the growth and flowering of plants. For example, continuous red light at 600 nm is effective in preventing flower formation. But if shortly after being exposed to this red light the plants are exposed to light of a longer wavelength at 730 nm, they will flower. The pigment responsible for this red and far-red effect is *phytochrome*. Phytochrome has two distinct forms, one with a maximum absorption in the *red* near 660 nm and another with a maximum absorption in the *far-red* near 730 nm (Figure 1.8).

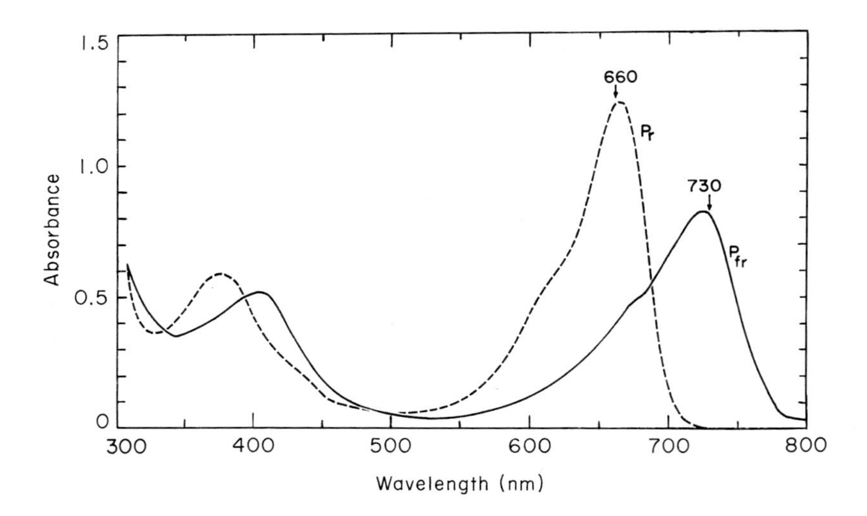

Figure 1.8. Phytochrome, absorption spectra of the red P_r form(— — — — —) and the far-red P_{fr} form (————), isolated from oats (Hartman, 1966, p. 352).

$$P_r \underset{730\text{ nm}}{\overset{660\text{ nm}}{\rightleftharpoons}} P_{fr} \xrightarrow{\text{darkness}} P_r$$

Its structure (Figure 1.9) resembles an open ring porphyrin (Figure 1.4), and with its system of conjugated double bonds a carotenoid molecule (Figures 1.5 and 1.6). The photochemistry of phytochrome (Figures 1.8 and 1.9) displays similarities to that of the visual pigment rhodopsin (Siegelman and Firer, 1964; Butler *et al.,* 1964; Hendricks, 1968).

There are many other pigment molecules associated with photobiological processes. Let us briefly mention a few of these pigments and their properties. The *cytochromes* are pigmented proteins that carry an iron atom in an attached chemical group. Their red color is derived from their prosthetic group or chromophore which is a derivative of iron protoporphyrin 9 (see Figure 1.4). The cytochromes are distinguished on the basis of their spectral absorption peaks in the reduced state and are designated by the letters a, b, c, and f. The spectra of reduced and oxidized cytochrome c are shown in Figure 1.10 and some comparative properties of the c-type cytochromes are listed in Table 1.1. The cytochromes function as electron carriers in reactions during photochemical processes.

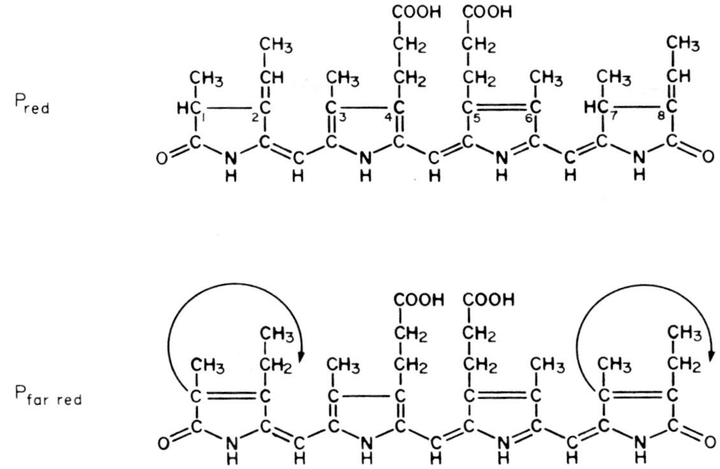

P_{red}

$P_{far\ red}$

Phytochrome

Figure 1.9. Phytochrome, chromophore molecular structures of the P_r and P_{fr} (Hendricks, 1968, p. 174).

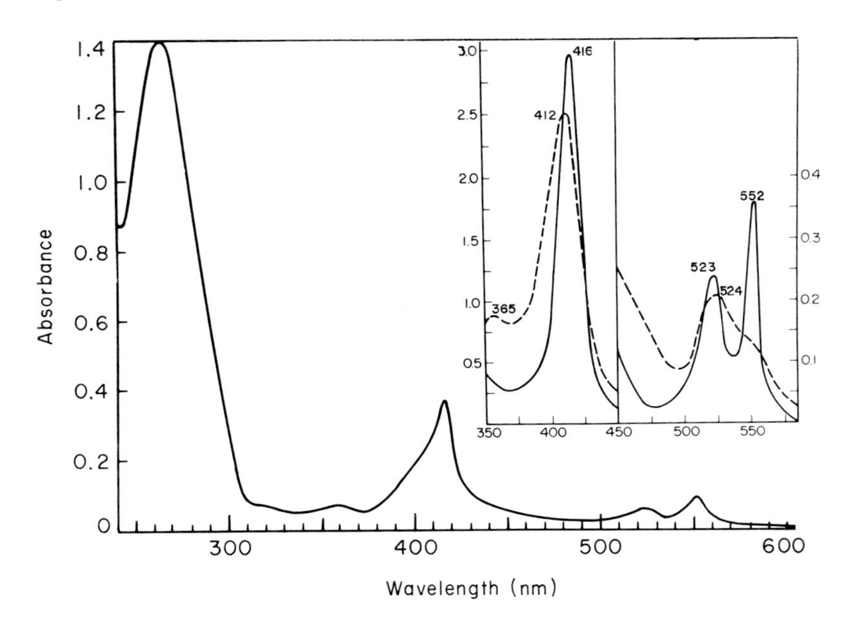

Figure 1.10. Cytochrome c absorption spectrum of the reduced state. Insert expanded to show the major reduced absorption peaks (————) and oxidized absorption peaks (– – – – –). Isolated from light-grown *Euglena gracilis.*

TABLE 1.1
COMPARATIVE PROPERTIES OF SOME C-TYPE CYTOCHROMES

	Bacteria[a,b]		Algae[c]		Higher plants[b]	Animal[b]
			Euglena gracilis			
	Rhodospirillum rubrum	*Chromatium*	Light-grown	dark-grown	Spinach	Beef heart
Absorption maximum in nm						
Oxidized						
α	535	525	524	530	535	535
γ	409	410	412	412	412	410
Reduced						
α	550	552	552	556	555	550
β	521	523	523	525	526	521
γ	416	416	416	421	417	416
Isoelectric point, pH	7.0	5.4	5.0	7.0	4.7	10.0
E_0' in volts, pH 7	+0.32–0.36	+0.01–0.04	+0.35–0.40	+0.31–0.33	+0.365–0.38	+0.26
Sedimentation, s_{20}	2.0	6.0	1.2	1.4	6.9	1.8
Molecular weight	16,000	97,000	11,000	13,000	110,000	13,600

[a]Bartch and Kamen (1960).
[b]Kamen (1956, 1960).
[c]Wolken and Gross (1963).

The *flavines* or flavoproteins, of which riboflavine, vitamin B_2, is an example, are a group of yellow photosensitive pigments. A molecule of riboflavine consists of the sugar, alcohol D-ribitol attached to a substituted isoalloxazine ring. Its structure and absorption spectrum are shown in Figure 1.11. Riboflavine upon ultraviolet excitation fluoresces green, 520–560 nm. It is light-sensitive in neutral or acid solutions, and bleaches to lumichrome with absorption peaks near 223, 260, and 360 nm. One of the more important chemical properties of riboflavine is its ability to change reversibly from the yellow-colored, oxidized form to the colorless, reduced form.

Ferredoxin is the name given an iron-protein which is neither a heme protein as are the cytochromes, nor a flavoprotein (Arnon, 1965). Unlike the cytochromes, which exhibit well-defined absorption peaks in the reduced state, ferredoxins have distinct absorption peaks in the oxidized state (Figure 1.12). It is also of interest to note the hypothetical scheme for photosynthesis as developed for the two light-absorbing chlorophyll systems (Figure 1.13). For, in photosynthesis, the initial photochemical

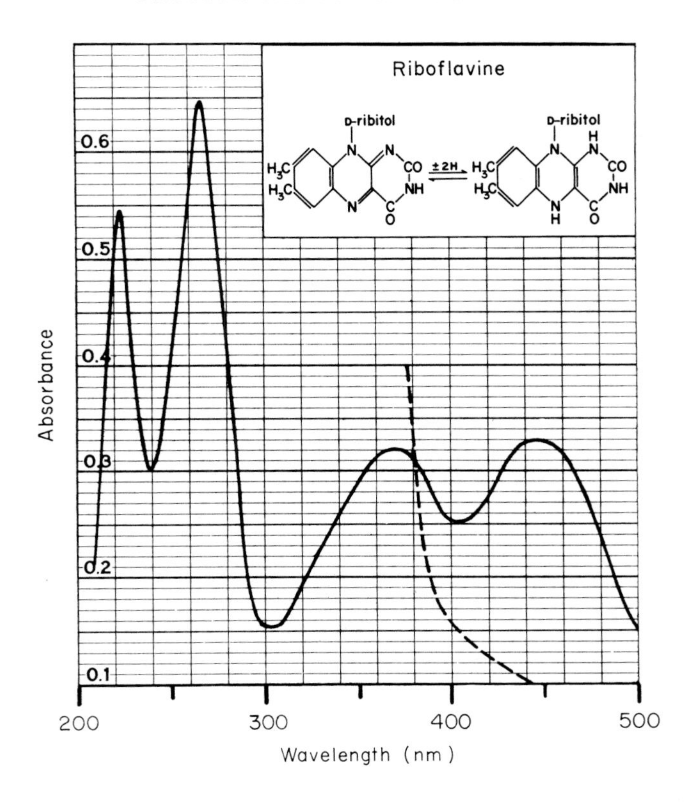

Figure 1.11. Riboflavine. The molecular structure and the absorption spectrum of the oxidized form, and when reduced with dithionite (— — — — —).

event is the absorption of light by chlorophyll a to produce an excited chlorophyll in which an electron is raised from its normal energy to a higher energy level. Such excited electrons flow from chlorophyll to ferredoxin to cytochromes via flavines and quinones. During this cyclic flow of electrons, the energy that the electrons initially acquired is funneled through oxidation–reduction reactions.

Melanin pigments are black, brown, or yellow. They occur in all animal phyla. They are synthesized along the pathway of tyrosine → dihydroxyphenylalanine (DOPA) → melanin.

Some properties of the carotenoids, flavines, and melanins are given in Table 1.2. These and other pigments, such as the pteridines and xanthommatins (Figure 6.19), will be discussed more specifically when considering how they are related to particular photobiological processes.

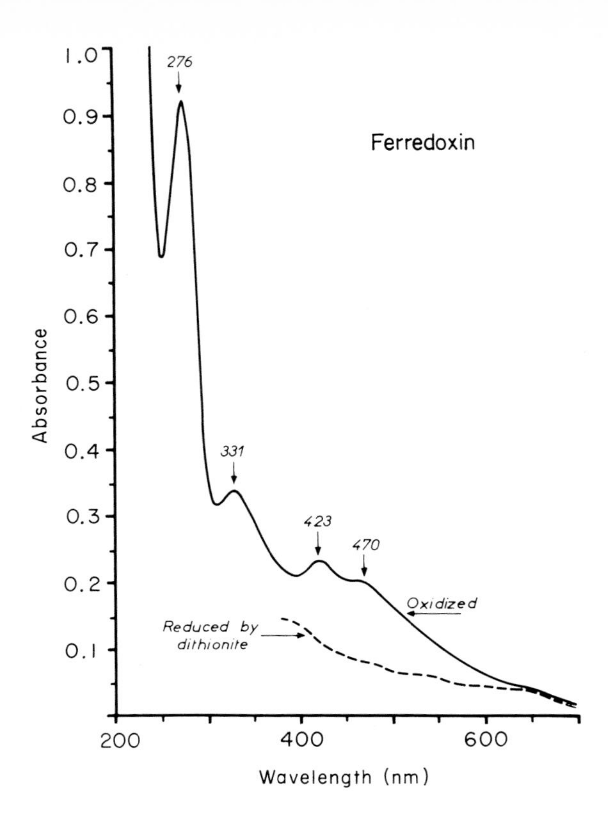

Figure 1.12. Ferredoxin. The absorption spectrum of the oxidized form, and when reduced with dithionite (– – – – –).

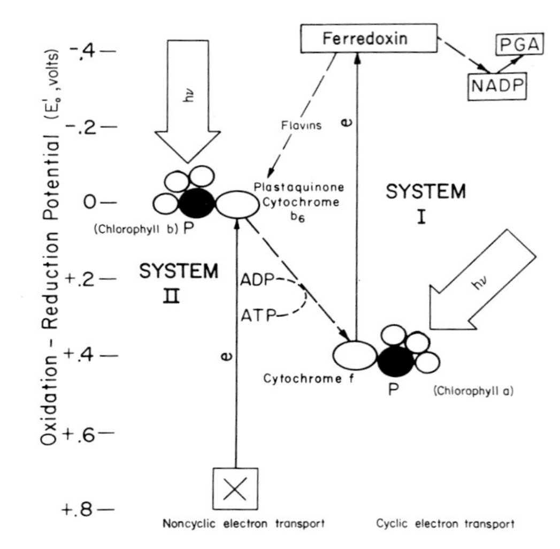

Figure 1.13. Photosynthesis, hypothetical scheme for the two chlorophyll systems, I and II.

TABLE 1.2
SOME PROPERTIES OF COMMON PIGMENTS[a]

Property	Carotenoids	Flavines	Melanins
Solubility			
Water	Generally no	Yes	No
Organic solvents	Yes	No	No
Color of solution	Yellow, orange, red	Yellow-orange	Yellowish, orange reddish, brown grey, black
Absorption bands, visible	Violet, blue, blue-green	Violet	None
Fluorescence	Carotene and lycopene, weak yellow-green; some carotenoid acids, blue	Green	None
Prosthetic group bound to	Protein generally	Protein	Protein (not always)
Behavior toward:			
Mineral acids	Very sensitive: conc., acids give blue or green colors, finally bleaching	Stable in dilute acids	Precipitated by dilute acids
Alkalis	Stable	Sensitive	Dissolved (or colloidal dispersion)
Oxidizing agents	Very sensitive, bleaching	Stable	Variable: decolorized by H_2O_2
Reducing agents	—	Reduced by $Na_2S_2O_4$ with bleaching, reversible	Reduce ammoniacal silver
Biological relations	Vitamin A	Vitamin B_2	Catabolite

[a]Taken in part from Fox (1953), p. 69.

Experimental Methods

Clues to how photoreceptors are structured to receive light stimuli and how they function can be found in their molecular anatomy and the molecular structure of their pigments. Much of the *fine structure* of photoreceptors was revealed by electron microscopy and by correlating these observations with data from light, polarizing, phase, and interference microscopy. With an electron microscope, resolutions close to 5 Å are possible, though for our work, resolutions near 20 Å were obtained.

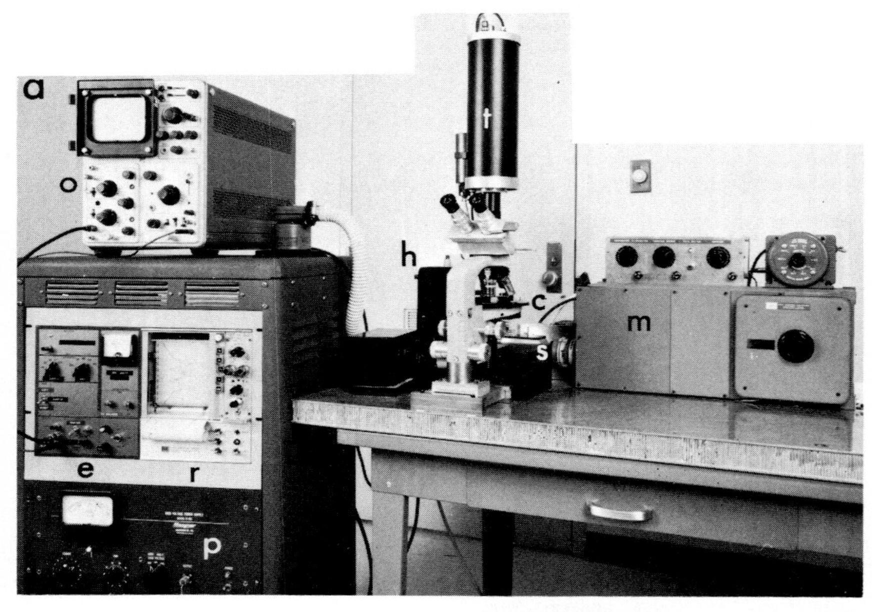

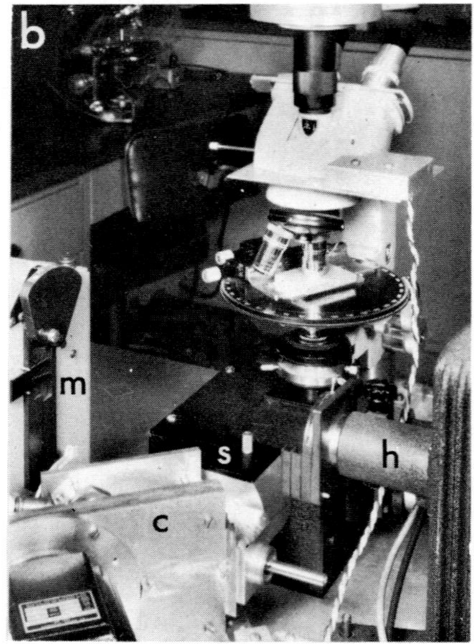

Figure 1.14. (a) Recording microspectrophotometer; (b) microscope stage, c, chopper; s, cuvette chamber; e, electronics, amplifier; h, lamp housing; m, monochrometer; o, oscilloscope; p, photomultiplier power supply; r, recorder; t, photomultiplier tube housing (Wolken *et al.,* 1968).

Hopefully, new methods in freeze-drying and freeze-etching for biological material, as well as technological developments for scanning electron microscopy will bring us closer to viewing biological structures in their natural state.

To identify the various pigments in the photoreceptors spectroscopic methods were employed. The systematic investigation of absorption spectra of pigment molecules was of key importance in establishing both their identity and structure. Compare, for example, in Figure 1.7, the absorption spectra of all-*trans*-β-carotene, retinal, and vitamin A_1 (refer also to Figures 1.2, 1.8, and 1.10–1.12).

To obtain absorption spectra by usual spectroscopic methods, known as solution spectroscopy, required extraction of tissue cells, usually with organic solvents; and in many cases further fractionization was required using differential solvent extraction and chromatography. In the 1930's Caspersson (1950) recognized that it was important to study the chemistry of cells in their natural state and began to develop microspectrophotometry for study of the cell nucleus. As a result, advances have been made in developing instrumentation for microspectrophotometry (Liebman, 1962, 1969; Wolken and Strother, 1963; Runge, 1966).

The microspectrophotometer designed and built in our laboratory is shown in Figure 1.14. This instrument is ideally suited for scanning living cells and recording the absorption spectrum of cell organelles approximately 0.5 μ in diameter. The response time of the instrument is of the order of milliseconds, and spectral measurements at low light levels from 230 nm in the ultraviolet through the visible to 800 nm in the infrared can be recorded in less than 2 seconds (Wolken *et al.,* 1968). It was used in these investigations to obtain the absorption spectra of *in situ* photoreceptors. From both solution spectroscopy and microspectrophotometry, relevant information was obtained regarding the nature of their pigment molecules.

In the following chapters, experiments will be described that were used to analyze the photobehavior and photoreceptor structure of numerous invertebrates, beginning with the protozoa. This will be followed by the structural analysis of invertebrate visual systems from a variety of insects, crustacea, and molluscs, together with the chemical isolation and identification of their visual pigments. A comparison will be drawn between these invertebrate studies and the vertebrate retinal photoreceptors—all in an attempt to push our understanding of photosensory systems to the limits of molecular biology.

II. THE PROTOZOAN PHOTORECEPTOR: EYESPOT AND FLAGELLUM

How a nerve comes to be sensitive to light, hardly concerns us more than how life itself originated; but I may remark that, as some of the lowest organisms, in which nerves cannot be detected, are capable of perceiving light, it does not seem impossible that certain sensitive elements in their *sarcode** should become aggregated and develop into nerves, endowed with this special sensibility.

Charles Darwin, 1859

Protozoa

How the first photoreceptor evolved is not known, but there is some basis for the suggestion that it may have been an adaptation by a cell for photosynthesis similar to that found in photosynthetic bacteria with chromatophores containing bacterial chlorophyll. Another speculation is that photoreceptors may have been the result of differentiation of flagellar processes of a cell in which photosensitive pigment globules became attached to or somehow became associated with a flagellum (Eakin, 1963). If so, this would provide a means for the cell to search for light of the appropriate intensity and wavelengths for photosynthesis. Examples of just such a close association can be seen in the protozoan flagellates; in *Chlamydomonas,* the eyespot is found within the chloroplast, and in *Euglena* the eyespot is associated with the flagellum.

In search of answers to how photoreceptor systems developed, how they are structured for light capture, and how they function as an integrated system, it seemed to me that some of the answers would be found among the protozoans that border between plants and animals, such as the flagellate, *Euglena* (Wolken, 1967).

To taxonomists, *Euglena* (Figure 2.1) have presented a problem in classification for they embody features commonly found in both plants and animals. In T. H. Huxley's definition, animals are denoted as those forms of life which require preformed organic molecules in their nutrition and which possess such characteristics as locomotion, flexible cell

*Protozoa, class Sarcodena.

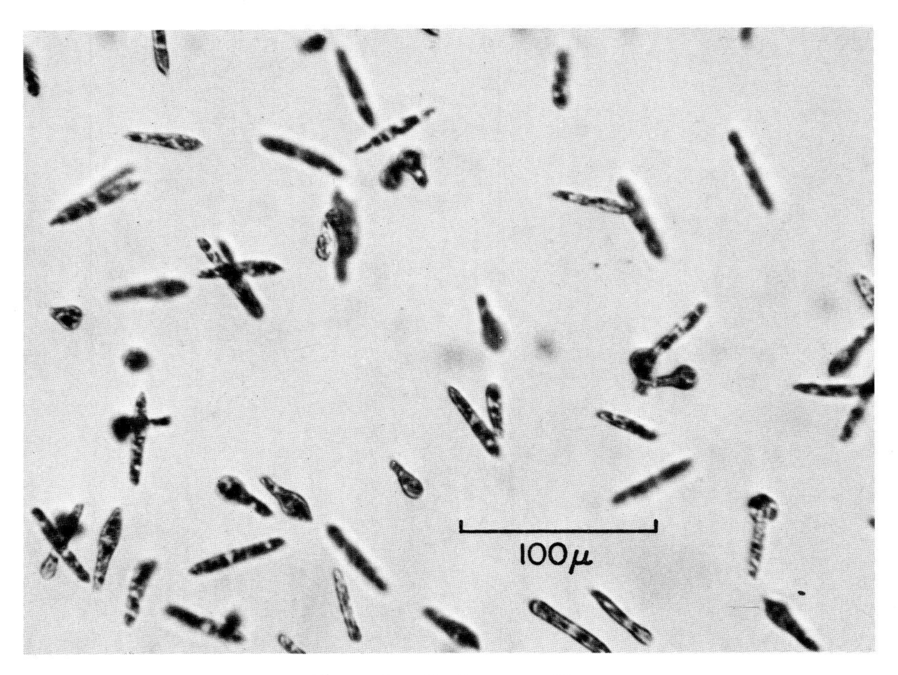

Figure 2.1. *Euglena gracilis.*

walls, and some kind of integrated "nervous" control. It is speculated that euglenas evolved close to the origins of the higher plants and most fungi. However, the euglenoid flagellates are probably remote from the protozoan flagellates that are supposed to have given rise to the cellular animals, i.e., the metazoa. The large number of question marks in Figure 2.2 indicates our primitive knowledge of the relationships among protozoan groups and their kinship to the metazoa. Hutner and Provasoli (1951) and Hutner (1955) look upon the algal flagellates as a heterogeneous group occupying a central position along the line of plant → animal descent.

Euglenas vary profoundly in structure and chemistry depending upon whether they live in light or in darkness (Figure 2.3). When living in the light they photosynthesize like a plant, but in darkness they lose their chloroplasts and hence their ability to photosynthesize. They then thrive by chemosynthesis, a process typical of all animal cells. The light (photosynthesis) ↔ dark (respiration) adaptation brings about changes in the chemistry and structure of the organism. This ability to adapt to light or darkness is fully reversible, providing mutations do not occur. Mutations

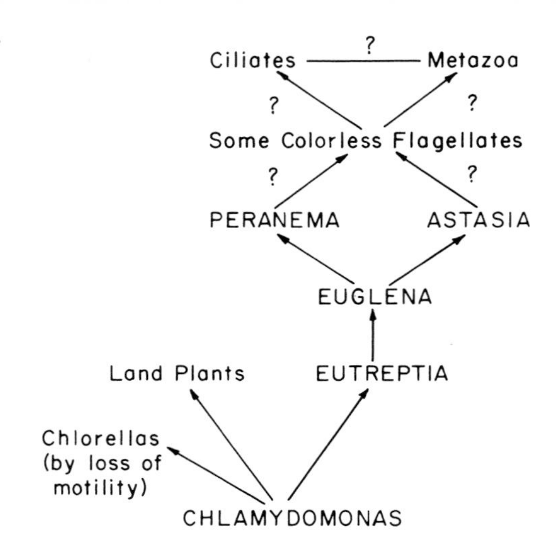

Figure 2.2. Affinities of the protozoan flagellates.

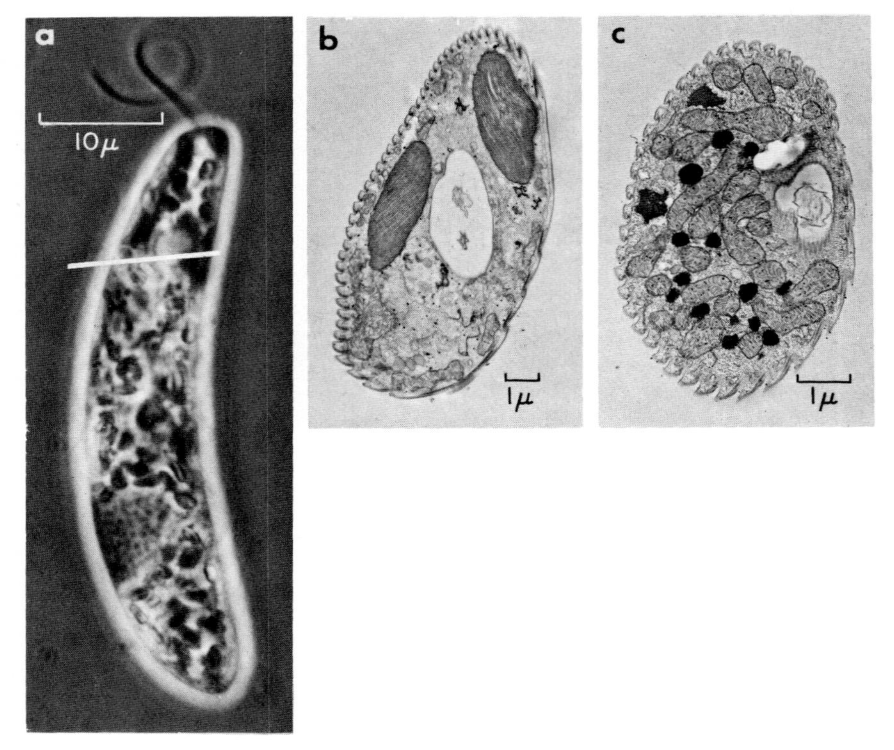

Figure 2.3. *Euglena gracilis.* (a) Light micrograph; (b) electron micrograph of cross section through the gullet area of light-grown *Euglena;* (c) electron micrograph of cross section through the same area of a dark-grown *Euglena.*

to nonphotosynthetic euglenas can easily be produced by changing the physical and chemical environment; for example, by growth temperatures above 33°C, by ultraviolet and ionizing radiation, and chemically by drugs such as streptomycin or antihistamines (Wolken, 1967).

Euglena gracilis

The photosynthetic *Euglena gracilis* is an elongated green cell from 50 to 100 μ in length and 15–30 μ in diameter (Figures 2.1 and 2.3a). The green color comes from the chlorophyll in its chloroplasts (Figures 2.3b and 2.4c). It has an exoskeleton or pellicle (Figure 2.4a) which permits the organism to expand or contract very much like a collapsible cup. From its gullet (cytopharynx) protrudes a flagellum, and closely associated with the flagellum is an orange-red eyespot (Figures 2.3a and 2.4d). Having both an eyespot and a flagellum, *Euglena* is provided with a photoreceptor system well suited for light searching.

FLAGELLUM

The flagellum is of the order of 30 μ in length. Attached to it are small lashlike fibrils called *mastigonemata* (Figure 2.5). In cross section the flagellum is from 0.25 to 0.40 μ in diameter and consists of a number of elementary filaments, *axonemata,* which are embedded in a matrix and covered by a membrane. The elementary filaments number eleven pairs, of which nine are peripherally located (Figure 2.6c,d), while the other two are found at the center of the flagellum. This arrangement is observed for a variety of plant and animal cells with flagella or cilia, from bacteria to the sperm tails of man (Fawcett and Porter, 1952, 1954; Manton, 1952; Pitelka, 1963, 1969; and Pitelka and Schooley, 1955).

EYESPOT

The eyespot, sometimes referred to as the *stigma,* is believed to be the primary region for light reception. It was speculated by Englemann (1882) that the eyespot is analogous in function to the retinal cells of the eye. Later, Mast (1911) suggested that a photosensitive pigment lay inside the eyespot area.

When sections of *Euglena* are observed with an electron microscope, the eyespot appears as an agglomeration of 40 to 50 pigmented granules varying in size from 0.1 to 0.3 μ in diameter, which forms a cross section of 2 $\times$ 3 μ (Figure 2.4d). The granules of the eyespot are located immediately below the membrane of the reservoir, a smooth-walled chamber that

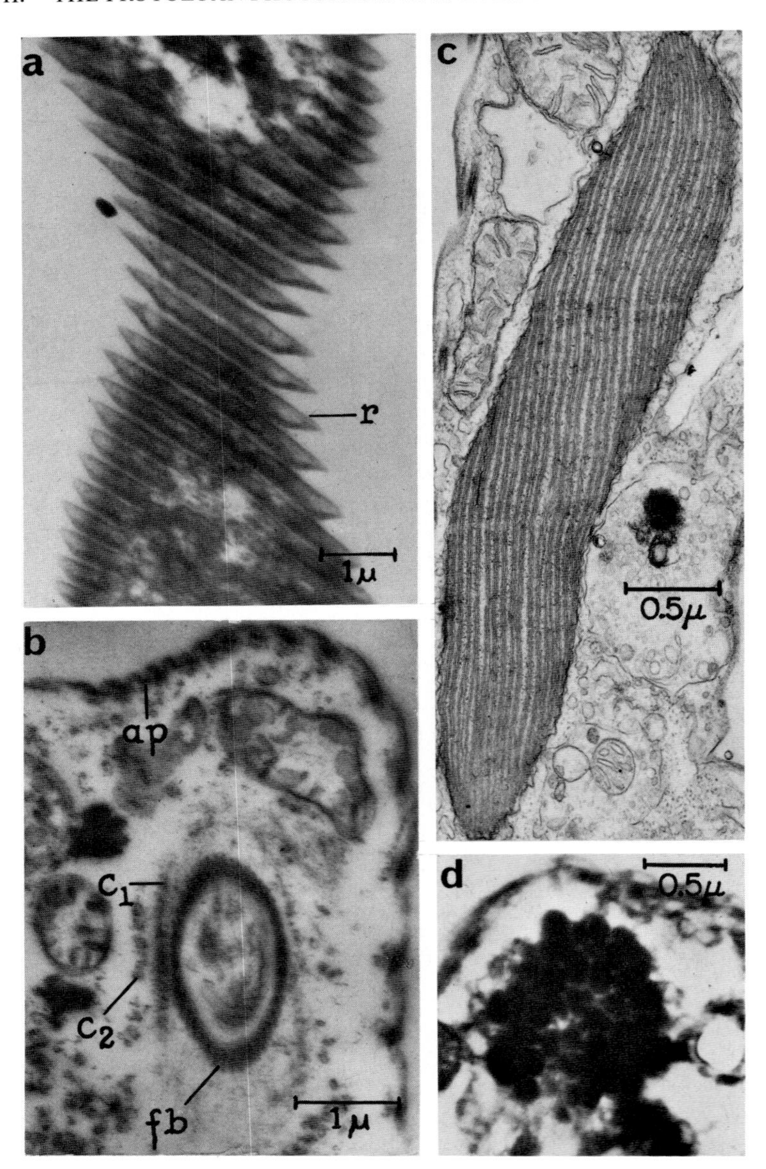

Figure 2.4. *Euglena gracilis.* Structures (a) pellicle or exoskeleton; r, structure of the macromolecules of the exoskeleton; (b) gullet area (fb. fibrillae in the wall of the gullet; c_1, network of very fine fibrils; c_2, element of the endoplasmic reticulum; ap, ridges of the pellicle, the same as r); (c) chloroplast; (d) eyespot granules.

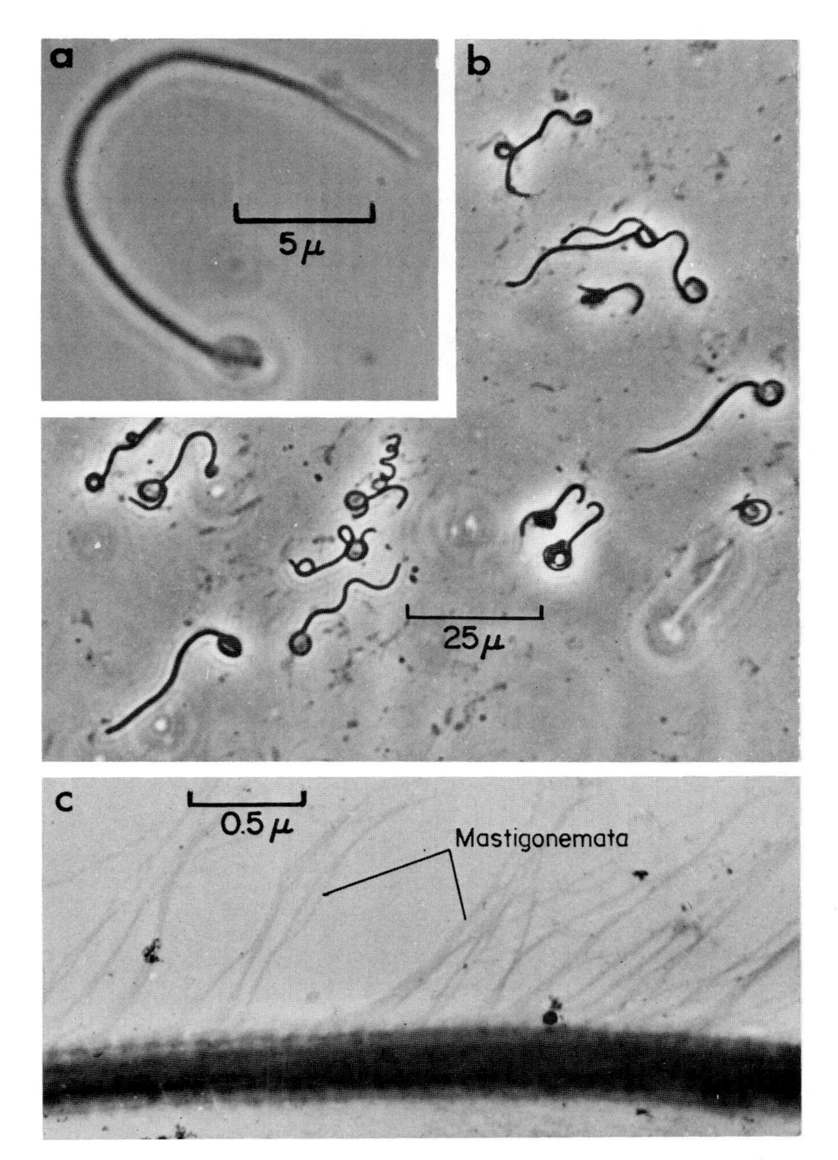

Figure 2.5. Flagella isolated from *Euglena gracilis*. (a), (b), and (c) at various magnifications.

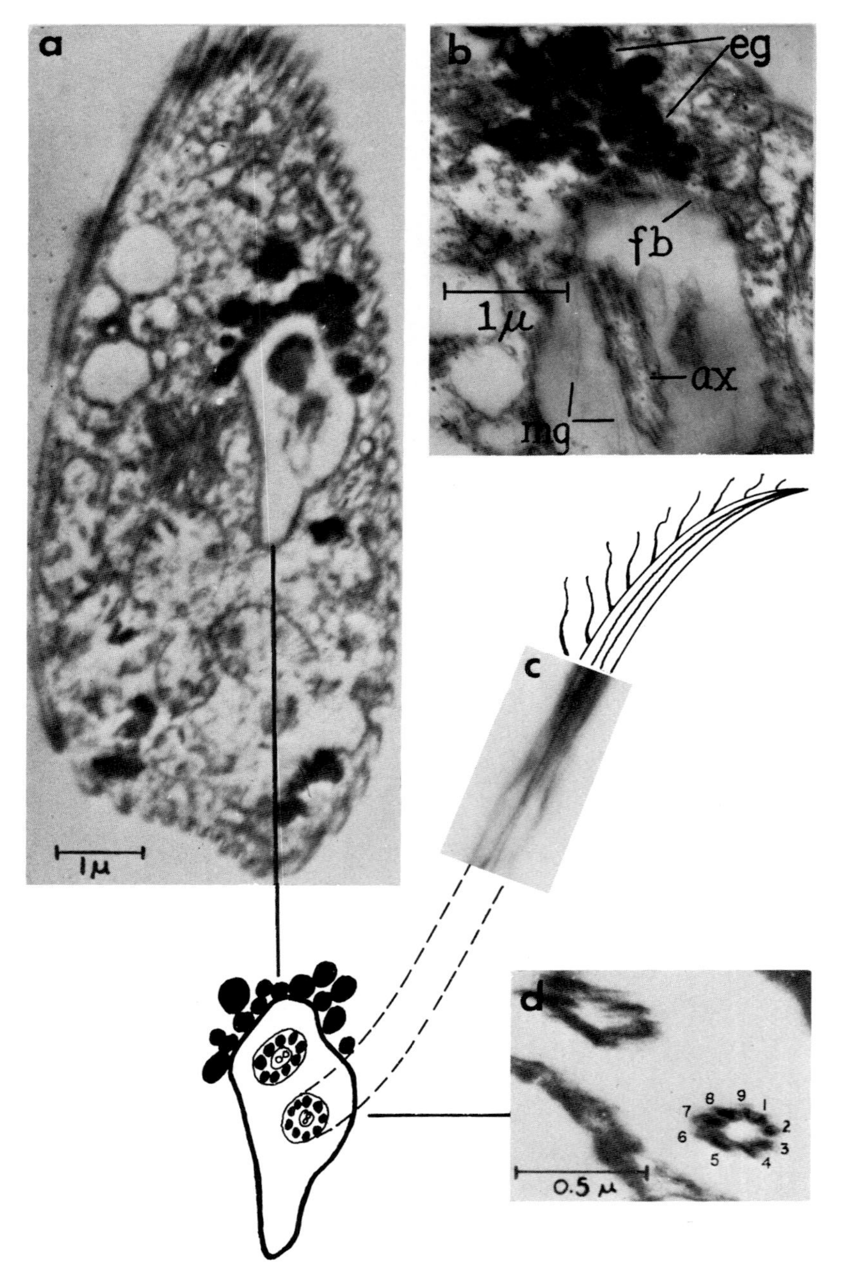

Figure 2.6. *Euglena gracilis,* structural relationship between the eyespot and the flagellum. (a) Cross section showing orientation of eyespot. (b) Eyespot granules, eg; fb, fibrillar system of gullet; ax, axonemata; mg, mastigonemata. (c) Longitudinal section of the flagellum. (d) cross section of flagellum, showing nine fibrils.

follows the ridged gullet in which the initial part of the flagellum is found (Figures 2.4b and 2.6). A system of regularly spaced fibrillae is found between the granules and the membrane of the reservoir (Figure 2.6b). Occasionally lenslike structures are also found near the eyespot (Wolken, 1967). A dense, homogeneous structure attached to the flagellum and facing the eyespot is identified as the *paraflagellar swelling* (Figure 2.7) and has long been considered the photoreceptor. If this is so, then the eyespot pigment granules would serve as a shading device or light filter (Wolken and Palade, 1953; Wolken, 1956a; Pitelka, 1963; Pringsheim, 1963).

Photomotion

To establish that the eyespot and flagellum are indeed *Euglena's* photoreceptor system for light searching it is necessary to determine its response to the direction, intensity and wavelength of the light stimulus.

Euglena reacts to light by swimming to or away from the light stimulus; thus it is phototactic. I have classified such phototactic behavior under the general term *photomotion,* and further separated the light response into *photokinesis,* defined here as the change in velocity or rate of swimming upon illumination without regard to orientation, and *phototaxis,* the directed orientation to light of specific wavelengths. Other photobehavioral studies of animals and definitions of these responses are described by Fraenkel and Gunn (1961).

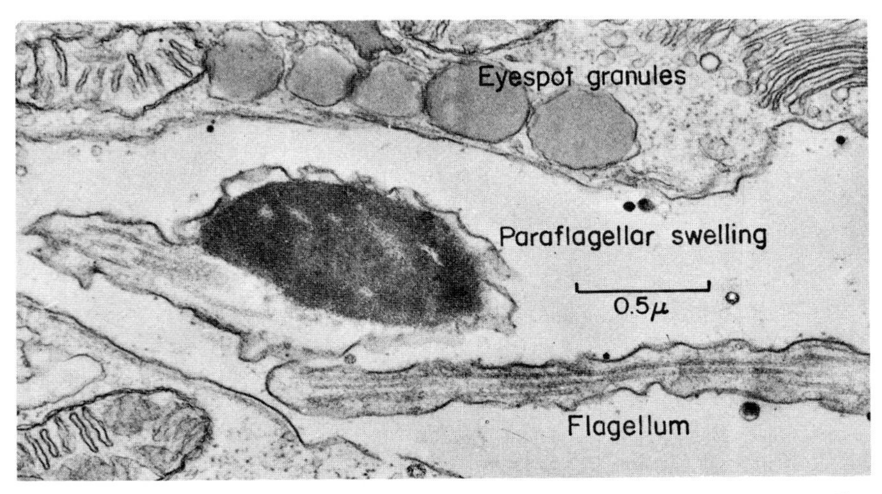

Figure 2.7. Eyespot granules and flagellum of *Euglena gracilis,* note paraflagellar swelling attached to flagellum.

There are, however, a number of important questions to be asked concerning phototactic behavior in general: What are the photoreceptor structures? What is the primary receptor pigment molecule? What is the fine structure of the photoreceptor? And with specific regard to *Euglena,* we must ask whether the hypothesized eyespot–flagellum system is not in fact the primitive counterpart of more highly evolved sensory systems.

The action spectrum for photokinesis and for phototaxis should indicate the absorption characteristics of the molecule responsible for such phototactic behavior.

PHOTOKINESIS

The rate of swimming of *Euglena* was measured by suspending the organism in culture media which had a viscosity of 0.987 centipoises (cp) at 25°C. The measurements were made with a microscope in a dark room at 25°C by timing the distance an individual organism moved in a calibrated, uniformly illuminated, Levy counting chamber.

The velocity of free forward swimming without any sideways rotation was measured by counting the number of squares on the chamber through which the organism swam and recording the time elapsed. A typical swimming pattern is shown in Figure 2.8. The average distance that *Euglena* traveled was found by plotting on graph paper the actual patterns of numerous organisms. To have statistical value, only those organisms which swam through 15 or more squares were recorded.

Experiments of this type were performed in white light, polarized white light, and with various wavelengths and light intensities. For the experiments with differing wavelengths, filter combinations were mounted at the light source, and the dominant wavelength of each filter was taken as the experimental wavelength. The relative-energy distribution curves and the corresponding dominant wavelengths of each filter combination are listed in Table 2.1. The light source was a General Electric 20 W daylight fluorescent tube, and the intensity was adjusted with neutral density filters. It was observed that the swimming velocity did not immediately change when a change in illumination was introduced to the organisms. It took 10 to 15 minutes for an observable, regular pattern of motion to give symmetrical distribution curves (Figure 2.9). A lag period of 10 to 15 minutes for a change in velocity to take place following a change in illumination was previously observed in *Volvox, Chlamydomonas,* and *Euglena* (Holmes, 1903; Mast, 1911; Loeb, 1918). A possible interpretation of this slow reaction is that it may depend upon a mechanism that is only indirectly affected by illumination.

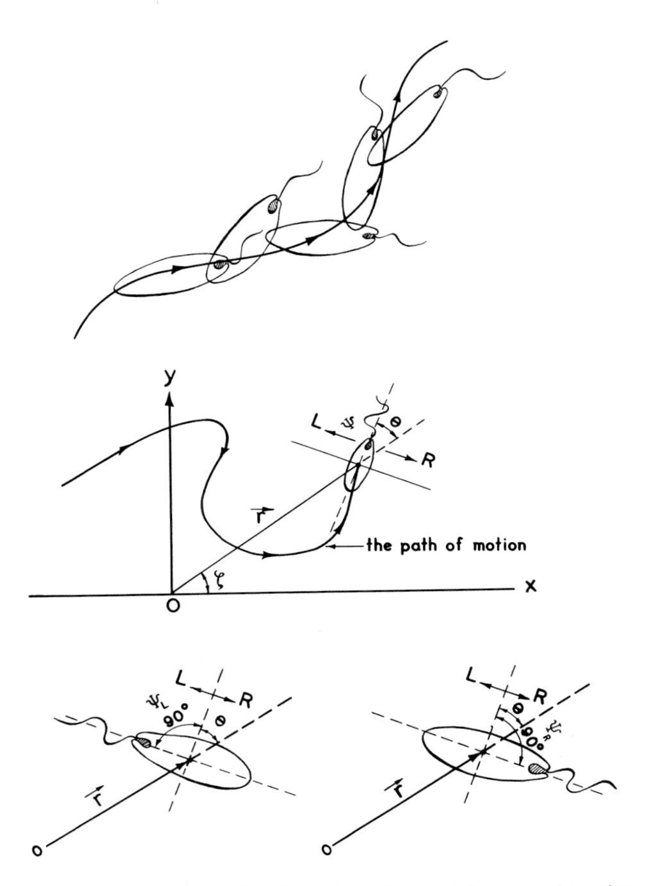

Figure 2.8. *Euglena,* swimming orientation and degrees of turning.

TABLE 2.1
FILTER COMBINATION USED TO DETERMINE PHOTOKINESIS ACTION SPECTRUM

Filter combination[a]	Dominant wavelength (nm)	Percentage transmission
5113 + 5120	418	11.0
3389 + 5113	442	5.8
3389 + 5562	462	38.0
3387 + 5562	480	26.0
3484 + 4303	545	15.0
2434 + 9780	600	18.0
2412	630	7.0
2434 + W-36	700	1.5

[a]Corning filter number, W = Wratten filter.

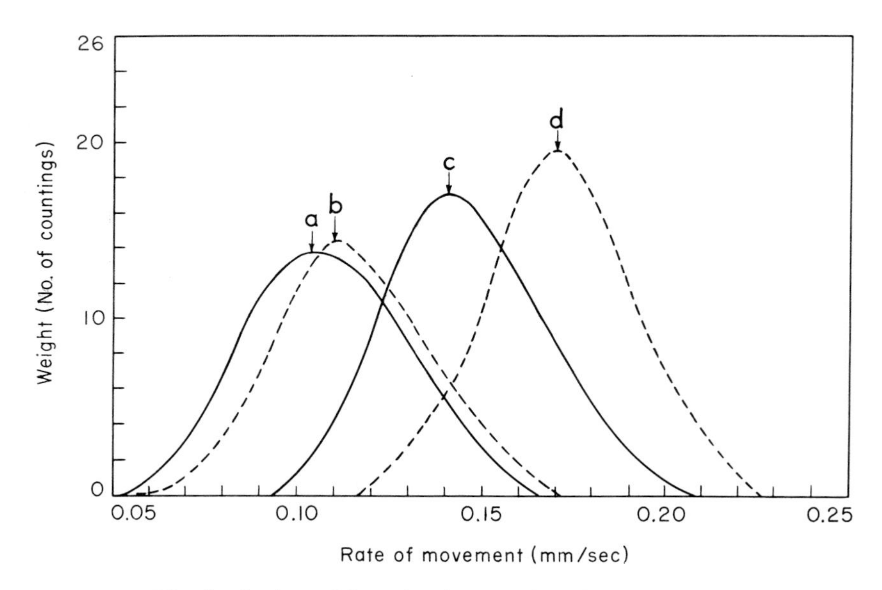

Figure 2.9. The distributions of the swimming rate of *Euglena* in polarized (- - -) and non-polarized (———) light. Plotted weight, number of countings versus velocity. The arrows indicate the centers of gravity. Note the shifts of the peaks. a, 4 fc of white light; b, 4 fc of polarized white light; c, 10 fc of white light; d, 7.5 fc of polarized white light.

The relationship between the velocity or rate of swimming and light intensity is presented in Figure 2.10a. It will be noted that the mean velocity rises sharply from 0.11 mm/sec at 2 foot-candles (fc) until it reaches a maximum rate of 0.16 mm/sec at 40 fc, the saturation intensity. It then starts to decrease slowly as the light intensity is raised above 40 fc to 150 fc. In polarized light, the swimming rate reaches a maximum average velocity of 0.18 mm/sec when the intensity of light is raised to 20 fc and then declines sharply with increased intensities (Figure 2.10b).

The significance of the saturation value is that the number of light quanta at this particular intensity is sufficient to excite all the molecules that have a thermal energy (measured by intensity of illumination and the absolute temperature) equal to or greater than the minimal thermal energy. Since the maximum absorption capacity has already been reached, beyond this intensity the absorption rate will remain constant and undisturbed by the extra number of quanta reaching the molecules. The same effect was observed for various single wavelengths, although the saturation intensity was different in each case.

In evaluating the relative effectiveness of various wavelengths on the

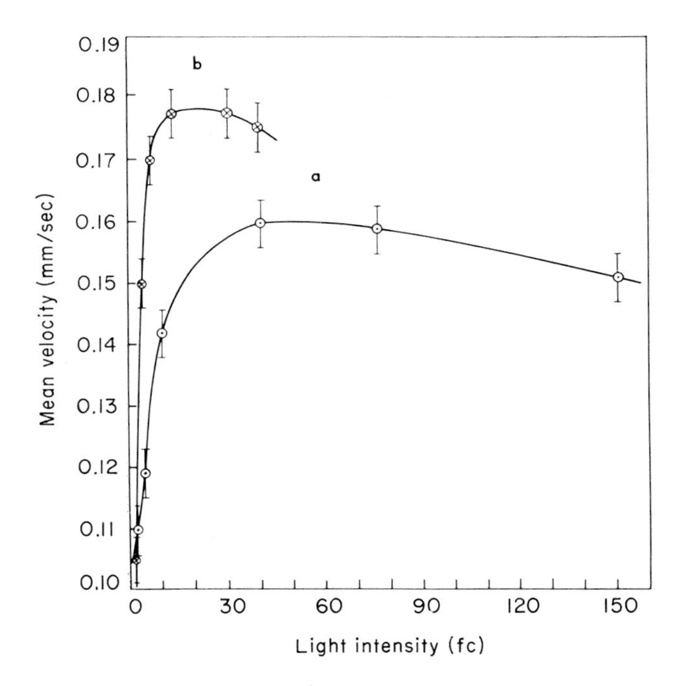

Figure 2.10. The mean velocity for *Euglena* swimming at various light intensities; a, in nonpolarized light; b, in polarized light.

rate of swimming, it was found that the mean velocity (in millimeters per second) versus light intensity was linear for light intensities less than 15 fc. The action spectrum plotted for the rate of swimming (mean velocity in millimeters per second versus wavelength at 4 fc) is illustrated in Figure 2.11. It will be observed that there is a major peak at 465 nm and another peak near 630 nm. This action spectrum is indicative of the absorption spectrum of the pigment involved in the rate of swimming. The peak at 465 nm suggests a carotenoid, whereas the peak at 630 nm would suggest another molecule.

The energy necessary to produce a response can be calculated from the area of the photoreceptor, the effective wavelength, and the light intensity. The estimated photokinetic energy for *Euglena* was 1.7×10^{-11} ergs/cm^2-sec. This calculation was made from the average cross section of the eyespot (6×10^{-8} cm^2), the wavelength peak at 465 nm (Figure 2.11) at 4 fc light intensity, and with the assumption that 10% of the radiation was absorbed. It is interesting that the number of photons that can excite the "eye" of *Euglena* at 465 nm is seven; the human eye can detect a minimum of four photons at 500 nm.

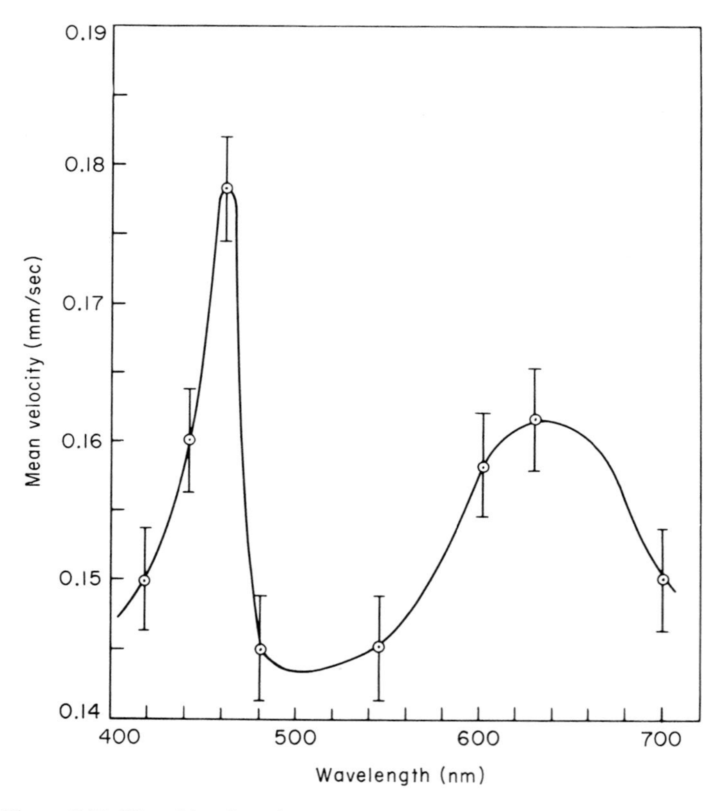

Figure 2.11. Photokinesis action spectrum for *Euglena* (measured at 4 fc).

Lee (1954a,b) studied the forward swimming of *Euglena gracilis* from 6°C to above 30°C. He found that the forward rate of swimming increases with temperatures, from 0.013 mm/sec at 6°C to a maximum of 0.08 mm/sec at 30°C (Figure 2.12). Above 30°C to 40°C forward swimming decreased to a rate of 0.038 mm/sec. It is possible from these data to estimate the energy necessary for forward movement. The minimum molar energy found by this approximation was 66 kcal/mole or 4.2×10^{-11} ergs/cm^2-sec of light energy. This figure compares favorably with the estimate from the photokinesis action spectrum through these two methods of calculating the excitation energy are independent of one another. Thus we see that the energy from light alone, if absorbed at the

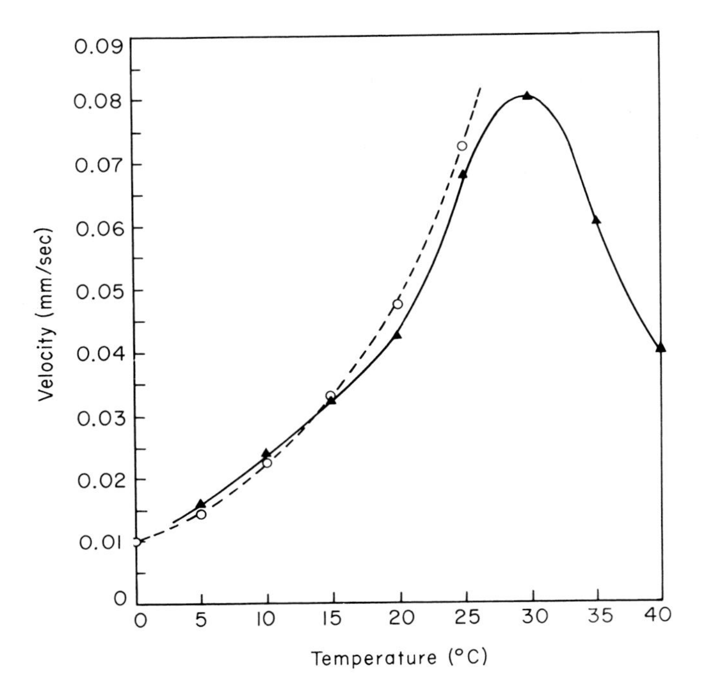

Figure 2.12. *Euglena* swimming velocity at various temperatures (Lee, 1954, p. 275). — — — o — — — computed curve (Wolken and Shin, 1958).

eyespot and transferred to the flagellum, is sufficient to cause the organism to move.

Phototaxis

The degree to which *Euglena* responds to different wavelengths was studied experimentally by obtaining the spectral sensitivity or phototaxis action spectrum (Mast, 1911; Loeb, 1918; Manten, 1948a,b; Clayton, 1953; Bünning and Schneiderhöhn, 1956; Batra and Tollin, 1964; Halldal, 1964; Diehn and Tollin, 1966).

To approximate the phototactic response of a population of *Euglena,* we dark-adapted the organisms and placed a known population into a long piece of glass tubing. Various filter combinations were inserted along the length of the tube and all were adjusted for equal light intensities. The tube was shaken and the organisms were permitted to swim freely to a colored filter of their choice. The accumulation, in terms of density of organisms, in front of each filter combination was determined at various time intervals. It was found that after 15 minutes a greater density of organisms accumulated in a narrow region near 465 nm and within a broader region from 570 to 600 nm. The major peak at 465 nm agreed well with that found in the photokinesis action spectrum (Figure 2.11).

For a quantitative study of phototaxis, a spectrophotometer was constructed, as illustrated in Figure 2.13. The apparatus was a completely

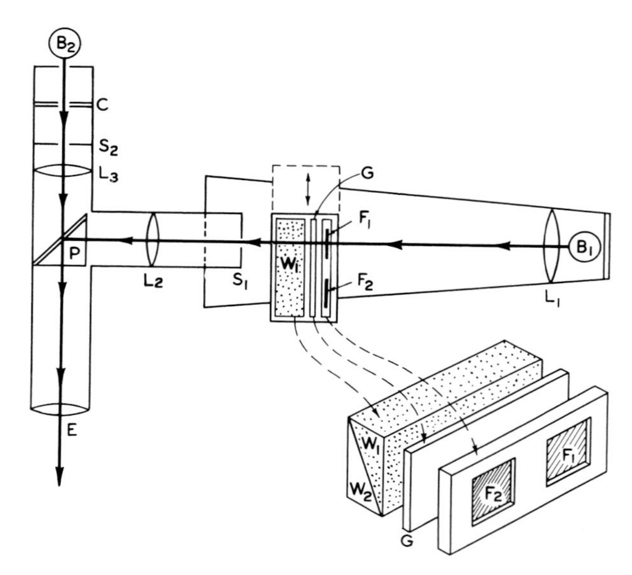

Figure 2.13. Apparatus for phototaxis measurements. L_1, L_2, L_3—Convex lenses; S_1, S_2—slits; B_1, B_2—light sources; W_1/W_2—experimental cell chamber; C—spectrophotometer scale; E—eyepiece; F_1, F_2—filters; G—glass plate; P—half-reflecting prism.

enclosed black box; the only light entering the test cell was from the light source B_2. The light from light source B_1 was collimated by the convex lens L_1 to the filter, F. The glass plates, G, were used to adjust the intensity of the filtered light passing through the experimental cell (W_1 and W_2). In these experiments a suspension of *Euglena* was placed in the cell compartment W_1, and the culture medium in W_2 as a reference. The light transmitted through both W_1 and W_2 passed through the slit S_1, the lens L_2, and was then reflected 90° at the half-reflecting prism P to the eyepiece, E. W_2 counterbalanced the deflection of the light passing through the prism formed by the organisms and medium inside W_1. The light from source B_2 passed through the scale plate C, the slit S_2, and the lens L_3, which projected the scale onto the eyepiece where it could be read.

The principle upon which this measurement is based is given by the equation:

$$I_0/I = Ae^{-kx}$$

where I_0 is the light intensity before entering W_1, I is the light intensity after passing a distance x through the suspension of organisms inside W_1 and k and A are constants characteristic of the organism and the medium, respectively.

Two filter combinations corresponding to two different dominant wavelengths, λ_1 and λ_2, can be placed simultaneously in the filter compartment F, so that one half of W_1 is illuminated with λ_1 and the other half with λ_2. If at any instant, the concentration of organisms in W_1 on the side of λ_1 is greater than that on the side of λ_2, the slit image will shorten in vertical length more on λ_1 than on λ_2 Thus the relative spectral sensitivity for the wavelength λ_1 is defined as greater than that for λ_2. Similarly, observation can be made for another filter combination corresponding to dominant wavelengths λ_2 and λ_3 (Wolken and Shin, 1958).

The advantage of using the wedge-shaped cell is that the light intensity transmitted is a negative exponential function of the concentration times the distance of transmission, and the nonuniform vertical distribution of the concentration of the organisms is counterbalanced by making the angle of the wedge-shaped compartment W_1 sufficiently large. This makes it possible to obtain a sharper boundary at the top of the slit image.

By this method the relative spectral sensitivities or the action spectrum for phototaxis was obtained for dark-grown, dark-adapted, and light-grown *Euglena*. The action spectrum for light-grown *Euglena* shows a major peak at 490 nm, with some sensitivity near 420 nm and indications of a rise beyond 600 nm (Figure 2.14a). For dark-adapted and dark-grown *Euglena,* similar spectral peaks were obtained. However, in polarized light there are maxima at 468 and 508 nm (Figure 2.14c). The polarized light effects may be indicative of the presence of at least two light-absorbing pigments. If there are two pigments, the resulting polarized light shift could be due to a mutual energetic interference between them. The maximum at 468 nm (Figure 2.14b) corresponds to the peak of the photokinesis action spectrum (Figure 2.11). The action spectrum for positive phototaxis obtained by Bünning and Schneiderhöhn (1956) shows major peaks near 480 and 495 nm with additional peaks near 425 and 450 nm.

There is no explanation at present for the response we found beyond 600 nm in the near-red part of the spectrum. This may be the result of

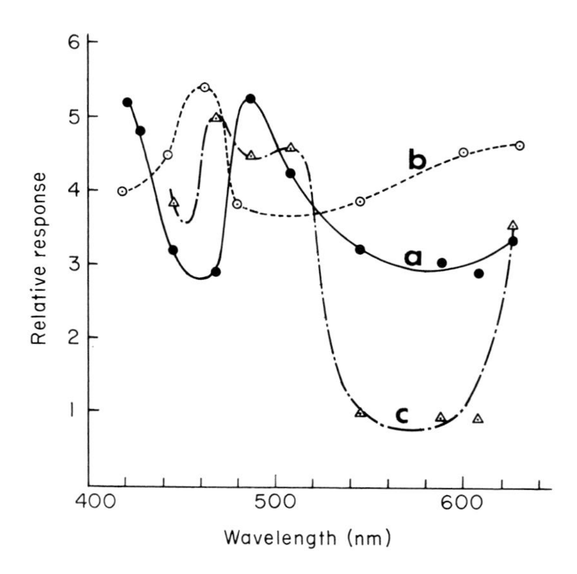

Figure 2.14. *Euglena gracilis* action spectra (a) ——●—— phototaxis; (b) ---⊙--- photokinesis; (c) ·——△——· phototaxis (in polarized light).

heat energy, energetic interference between two pigments, or the presence of other light-absorbing molecules within the organism. These results indicate that whatever pigments are responsible for phototaxis, they are present in light-grown, dark-grown, and dark-adapted euglena.

Photoreceptor Pigment

It has been generally assumed that the eyespot pigment is astaxanthin, mainly because of the discovery of astaxanthin in the cytoplasm of the red *Euglena sanguinea*. The pigment astaxanthin, 3,3′-dihydroxy-4,4′-diketo-β-carotene (Figure 1.6), has been encountered thus far only in animal tissue, in the eyes and in the integuments of crustacea (familiar to us as the red pigment of the boiled lobster), and as a screening pigment in the retinas of certain animals. The analysis by Tischer (1936) of the red pigment haematochrome, from *Euglena heliorubescens* and from *Haematococcus pluvialis,* demonstrated that the principal component of haematochrome is euglenarhodon ($C_{40}H_{48}O_4$), a ketonic xanthophyll. Euglenarhodon is also believed to be the red pigment of *Euglena rubra* (Johnson and Jahn, 1942). This pigment resembles, or is identical to, astacene (which has also been isolated from crustacea). Astacene is a ketonic carotenoid thought to be a degradation product of astaxanthin (Tischer, 1936).

There is no evidence to suggest that astaxanthin participates either in the visual chemistry of the retina or in the metabolism of vitamin A. The absorption spectrum of crustacean astaxanthin, unlike those of the common plant carotenoids, is a single broad band, maximal at about 500 nm in the blue-green, sloping off to about half that optical density at 460 and 560 nm (Goodwin, 1952).

What, then, is the eyespot pigment and the photoreceptor molecule? *Euglena gracilis* synthesizes three main carotenoids: β-carotene, lutein, and neoxanthin (Tables 2.2 and 2.3). Lutein was found to be the major pigment comprising 80% of the total (Goodwin and Jamikorn, 1954). The two pigments which persist in dark-grown *Euglenas* were found to be lutein and β-carotene.

TABLE 2.2
SEPARATION OF *Euglena gracilis* CAROTENOIDS BY COLUMN CHROMATOGRAPHY[a,b]

Zone number	Description	Spectral absorption maxima in light petroleum ether in nm	Identification
I	Brown-khaki	422, 448, 475	β-carotene
II	Lemon yellow	419, 442, 469	Lutein
III	Yellow	415, 438, 463	Neoxanthin

[a]Goodwin and Jamikorn (1954).
[b]Adsorbent: weakened alumina; developer: light petroleum ether containing different amounts of diethylether.

TABLE 2.3
THE CAROTENOIDS IN TWO VARIETIES OF *Euglena gracilis*[a]

Pigment	var. *bacillaris*[b] (%)	var. *fuscopunctata* (%)
β-carotene	11	15
Lutein	82	16
Neoxanthin	7	21

[a]Goodwin and Gross (1958).
[b]Seven days' growth.

Krinsky and Goldsmith (1960) attempted to identify the *Euglena* carotenoids and the eyespot pigment by chromatography. In their analysis they did not detect astaxanthin or astacene, but found 80% of the carotenoids to be antheraxanthin instead of lutein, 11% β-carotene, 7% neoxanthin, a small amount of γ-carotene, crytoxanthin, echinenone, and two previously unreported ketocarotenoids, euglenanone and hydroxyechinenone. In addition, Krinsky (1964) studied the carotenoid pigments following illumination under aerobic and anaerobic conditions, and found that under anaerobic conditions a photochemical conversion occurred from epoxide to nonepoxide carotenoids. This reaction was reversed in the dark and could be inhibited by aerobiosis, indicated as follows:

$$\text{Antheraxanthin} \underset{\text{dark}}{\overset{\text{light, } N_2}{\rightleftharpoons}} \text{Zeaxanthin}$$

Both the carotenes and neoxanthin remained essentially unchanged under these conditions (Bamji and Krinsky, 1965).

Assuming that the eyespot is a mixture of carotenoids, for example β-carotene, lutein, and neoxanthin, when dissolved in light petroleum ether, the absorption peaks should lie within the range 415 to 475 nm. When dissolved in carbon disulfide, a nonpolar solvent, the absorption peaks range from 450 to 510 nm. These and similar carotenoids could therefore account for the absorption peaks in the 450 to 510 nm region as found in the action spectra (Figure 2.14).

To circumvent the difficulties of isolating the eyespot granules and extraction, the microspectrophotometer was used to obtain *in situ* absorption spectra from single *Euglena* eyespots of 2 μ^2 areas. These data showed that the eyespot has a broad absorption band from 440 to 520 nm (Gössel, 1957; Strother and Wolken, 1960a, 1961; Wolken, 1967). More recent studies of absorption spectra of the eyespot area show that the peaks lie near 430, 465, and 495 nm, and near 350 and 270 nm in the ultraviolet (Figure 2.15a); spectra closer to the base of the flagellum (near the paraflagellar body) show absorption peaks about 440 and 490 nm (Figure 2.15b); the flagella do not show any peaks in the visible (Figure 2.15c). In the heat-bleached *Euglena* mutant which lacks chloroplasts, the eyespot spectrum shows absorption peaks near 430, 465, and 510 nm, and in the ultraviolet at 340 nm (Figure 2.16a). The eyespot spectrum (Figure 2.16b) resembles that of a flavine semiquinone (Land and Swallow, 1969).

When light-grown euglena are dark-adapted for about 1 hour and mounted on the cold stage (5°C) of the microspectrophotometer, the absorption spectrum obtained (Figure 2.17a) is similar to that found for

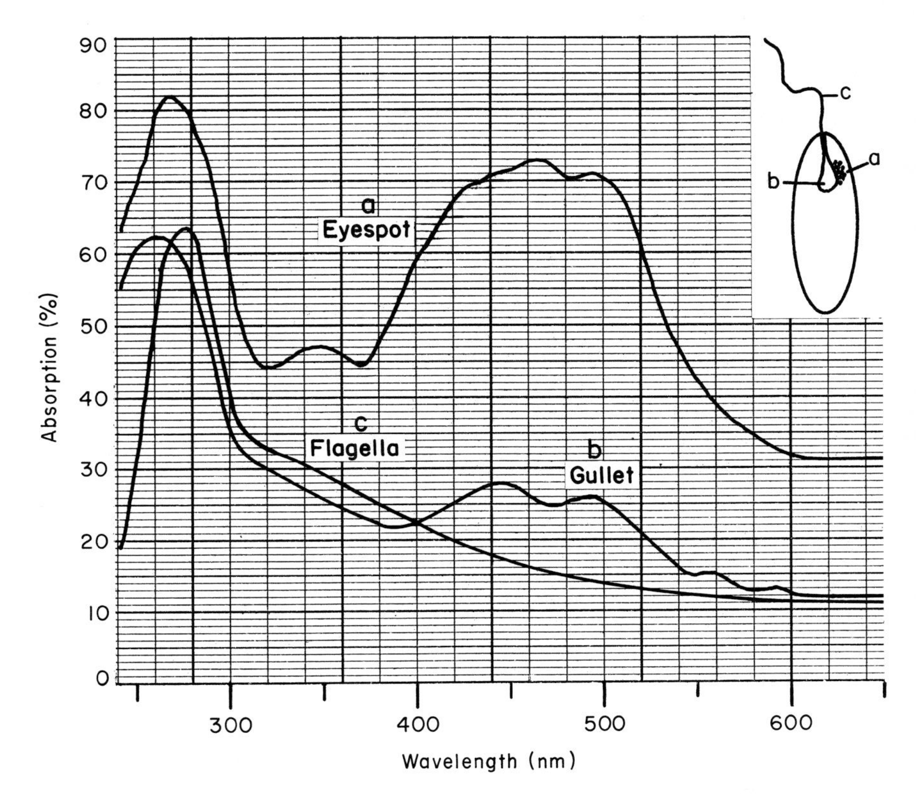

Figure 2.15. Absorption spectra, obtained by microspectrophotometry, of (a) eyespot, (b) gullet, and (c) flagella. Light-grown *Euglena gracilis*.

the eyespot in Figures 2.15a and 2.16a. When the same area is illuminated with strong white light for 1–5 minutes, the major absorption peak around 490 nm bleaches, shifting its major peak to 440 nm (Figure 2.17b). This absorption peak at 490 nm is close to the peak found for the phototaxis action spectrum. But what may be even more interesting is the similarity to bleaching of the visual pigment rhodopsin (discussed in Chapter V and in Chapter VI for the invertebrate visual pigments).

Interpretation of the spectra to establish the identity of the photoreceptor pigments in the eyespot is, however, extremely difficult. Tollin and Robinson (1969) suggest from their action spectra that the eyespot is a shading device for a photosensitive region located at the base of the flagellum and that the photosensitive pigment resembles a flavoprotein with absorption peaks about 460 and 370 nm (Figure 2.18b; compare with

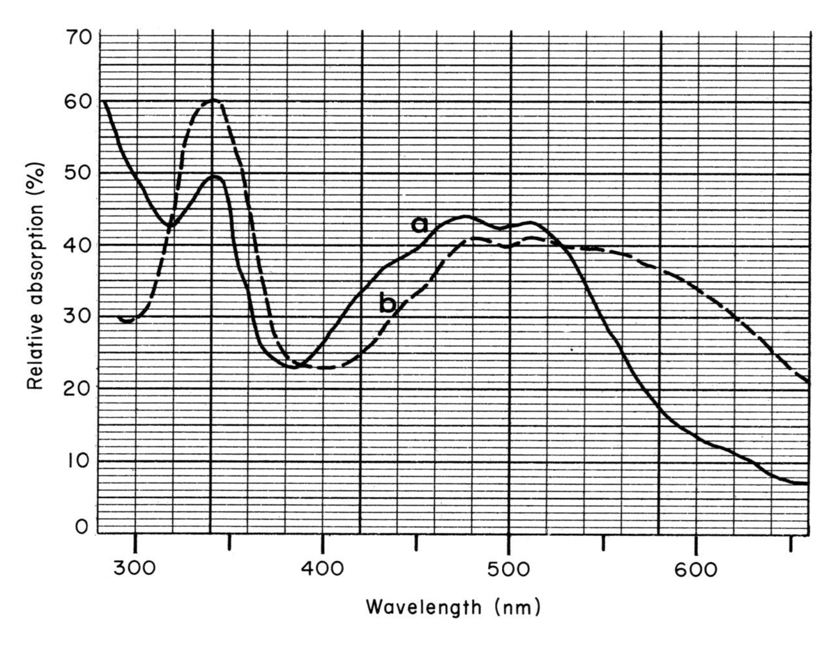

Figure 2.16. Eyespot absorption spectrum (a) of heat-bleached *Euglena,* obtained by microspectrophotometry, compared to (b) flavine semiquinone, pH 5.1 (from Land and Swallow, 1969, p. 2121).

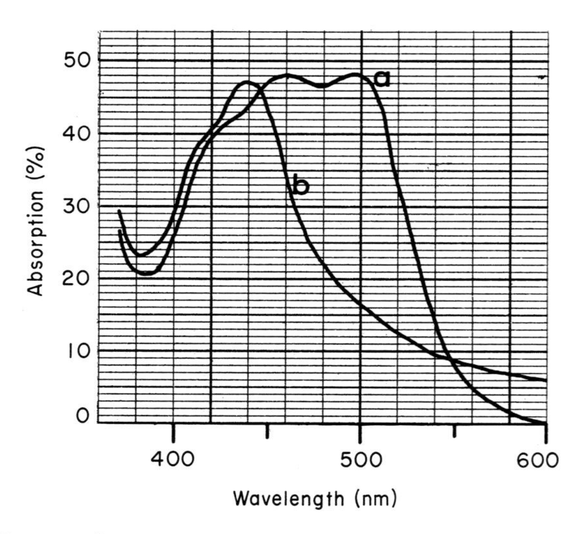

Figure 2.17. Eyespot absorption spectra obtained by microspectrophotometry (a) after dark-adapted for 1 hour and (b) the same eyespot after 5 minutes of white light.

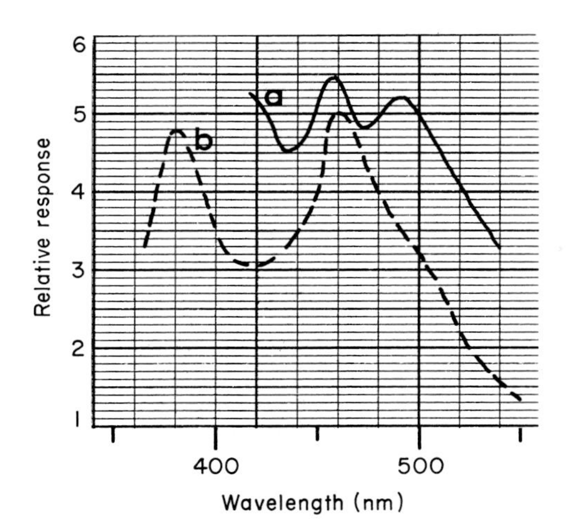

Figure 2.18. (a) phototactic action spectrum for *Euglena gracilis;* (b) action spectrum for photosuppression of phototaxis in *Euglena* (Tollin and Robinson, 1969, p. 415).

Figure 1.11). If a flavoprotein is the photoreceptor molecule, then an analysis of the *Euglena* flavines would give us some idea of its concentration. Analysis of the total flavines showed that for both light- and dark-grown *Euglena* there was of the order of 10^{12} flavine molecules per cell. This is more than sufficient when compared with the number of visual pigment molecules in the photoreceptors of all eyes, which contain from 10^6 to 10^9 rhodopsin molecules. However, the action spectra (Figures 2.11, 2.14, and 2.18) and the absorption spectra (Figures 2.15 and 2.16a) of the eyespot do not rule out that a carotenoid may be the photoreceptor molecule. On the assumption that the surface of the area occupied by the eyespot contains a monolayer of carotenoid molecules, e.g., β-carotene, 10^6 pigment molecules would fit around the eyespot. There is also the possibility that there are two pigments, a carotenoid and a flavoprotein, in which one is the primary photoreceptor pigment and the other an accessory or screening pigment in the process.

The Flagellum and Excitation

Electron microscopy of the flagellum shows that there are nine fibrils, each wound in a helical pattern with two central filaments. The *Euglena* flagellum also consists of what appear to be numerous segments along its entire length (Figure 2.5c). The base of the flagellum is situated in the vacuole very close to the eyespot (Figures 2.6 and 2.7). However, it is not known whether the flagellum is attached in some way to the eyespot, and how the excitation is carried from the light-absorbing pigment within the eyespot area to the flagellum.

The structure and chemistry of the algal flagella that have been studied indicate that they are composed almost entirely of protein (Lewin, 1955, 1962; Astbury and Saha, 1953). It has been found that the flagellar protein has similarities to myosin, the contractile protein of muscle (Lewin, 1962). Flagellar protein was prepared by Weibul (1951) which on hydrolysis yielded a mixture of amino acids, but the purified protein differed from muscle proteins in that it was deficient in the sulfur-containing amino acid, cysteine. Large quantities of flagella have been isolated from *Euglena* by continuous-flow centrifugation techniques. The absorption spectrum of these flagella show only one absorption peak, around 280 nm, typical of a protein (Figure 2.15c). These isolated flagella (Figure 2.5) display vigorous beating in 10^{-3} M solutions of ATP (adenosine triphosphate). It would be extremely interesting to know more about their biosynthesis, chemistry, and molecular structure, in particular about their matrix and sheath (Holwill, 1966; Rosenbaum and Child, 1967). If indeed the *Euglena* cell has a mechanism of excitation analogous to that of a nerve cell, it should possess the chemistry of a nerve cell. Experimental analysis does show that there are 3.85×10^2 acetylcholine molecules per cell, a compound which is associated with the chemistry of nerve excitation.

Flagellar Motion

The whipping action of the flagellum is probably adjusted so that the organism can move forward or backward and faster or slower. The light being absorbed by the pigment within the eyespot area would act as a continuous energy source for the motion. The energy flow itself is assumed to be discontinuous and supplied to the flagellum in unit pulses. The fact that the motion of the flagellum is very smooth indicates that the full length of the flagellum forms a certain regular pattern of motion. It should be noted that one whipping is not caused by a single pulse but is caused by many small pulses firing successively along the length of the flagellum.

Although no one has understood how the whipping causes the organism to perform with such a streamlined motion, it was suggested that the nine outside fibrils are perhaps wound helically around the two continuous central ones. However, electron micrographs do not show this. The flagellum was first thought of as a simple propeller. In Figure 2.19 the propeller is rotating clockwise, and the direction of the flow of the medium is shown

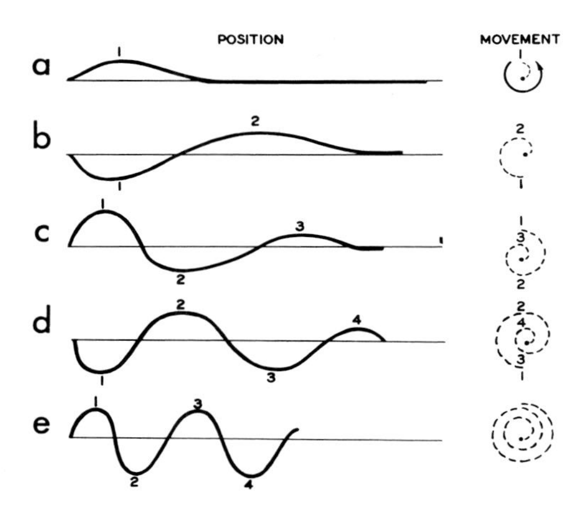

Figure 2.19. Possible whipping motions of flagellum, a, b, c, d, and e, position and movement.

by the arrow. Thus, the medium flows in the direction opposite that of the motion of the beat of the propeller. The surface of the propeller is so inclined that it pushes the medium to the back when it rotates clockwise, and forward when it rotates counterclockwise. If the nine fibrils were wound helically with a constant advancing period, and if the whipping were made by twisting the flagellum clockwise, the medium in which the organism swims would move forward, or if twisted counterclockwise, backward. The velocity would depend on the strength of the thrust, the number of whippings per unit time, and the distribution of impulse firings along the length of the flagellum. To keep the whipping action going at a constant rate, it is necessary that the thrusts occur at regular intervals along the length of the flagellum and at a certain critical instant. Also, if a thrust is given at a point to the right, the next one should be given to the left to keep the motion at a constant whipping radius. Therefore, it is conceivable that there is a feedback mechanism at work which lets the pulse firings occur at the side where the external mechanical stimulus is greatest.

By the same analogy, the two central fibrils are thought to be energy pulse carriers as well. One of them is designated the main pulse carrier and the other the feedback pulse carrier. The nine outside fibrils fire the pulses at various junctions along the flagellum, while the central fibrils bring the pulses to the junction. The direction of transfer to the main pulse carrier is from the eyespot to the flagellum, and can be controlled by the external light conditions.

This flagellar model permits us to conceive of a machine which theoretically could perform exactly the same kind of motion as *Euglena* upon stimulation with the light. One possible machine of this type is a photocell such as the one schematically drawn in Figure 2.20. In the diagram, P_1,

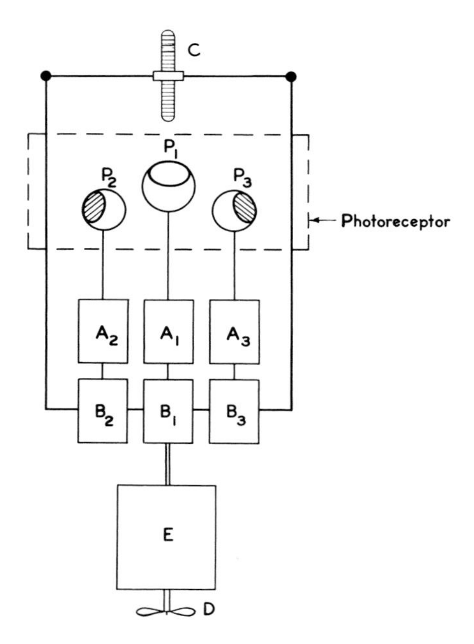

Figure 2.20. An analog "machine" of the photoreceptor system, that performs the same kind of light searching as *Euglena*. See text for explanation of letters.

P_2, and P_3 are photocells, of which P_2 and P_3 are the phototactic receptors effective in determining directional preference, while P_1 controls the speed of motion produced by the engine, E, and the propeller, D. A_1, A_2, and A_3 are the proper amplifiers which amplify the signals conveyed from P_1, P_2, and P_3 to B_1, B_2, and B_3, respectively. B_1, B_2, and B_3 are devices to "collect" the signals from P_1, P_2, and P_3, which results in mechanical motion at D and C. It is supposed that the photosensitive pigment in P_1 has a characteristic spectrum like that of photokinesis (Figure 2.11), while P_2 and P_3 contain a photosensitive pigment which has a characteristic spectrum like that of phototaxis (Figure 2.14a).

The Eyespot–Flagellum Photoreceptor System

Finally, it is of interest to consider *Euglena* as a primitive photosensory cell (Figure 2.21) and to examine more closely the eyespot and

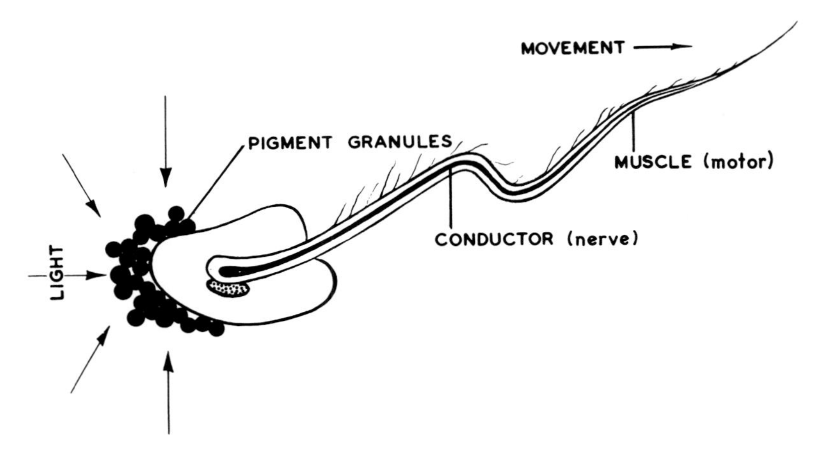

Figure 2.21. Photosensory cell, based on the eyespot-flagellum system.

the flagellum as the organism's device for nervous control. As already mentioned, in *Euglena* the eyespot (receptor) is intimately associated with the flagellum (effector and conductor). By means of the intensity and wavelength of light we can "communicate" with the organism to the extent that its speed and direction of motion are controlled. This in turn suggests a sensory cell or its analog, a photocell. *Euglena's* eyespot–flagellum system may therefore be regarded as a servo or feedback mechanism which endeavors to maintain an optimal level of illumination on the organism for light capture and photosynthesis. The eyespot and flagellum are somehow linked so that light falling on the "eye" produces motion. This translation of an internal effect into a surface action produces problems similar to those involved in the origin of nervous impulses in animal photoreceptors. Similar hypotheses have also been formulated concerning the relations among phototropism, phototaxis, and vision in higher animals (Wolken, 1967; Delbrück and Reichart, 1956; Clayton, 1953). Structurally the most interesting features in the development of the retinal photoreceptors—the rods and cones of the eye—are the flagellum-like fibrils which connect the inner and outer segments (Figure 5.5). Willmer (1955) suggested that embryologically the retinal rods and cones arise from flagellum-like processes. The evidence of Sjöstrand (1949, 1953a, 1953b) from electron microscopy shows that the connection consists of a bundle of fibrils arranged in a pattern not unlike what is observed in the *Euglena* flagellum (Figure 2.6). In the chrysomonad *Chromulina* there is a second flagellum associated with the eyespot area which possesses a fine structure of lamellae that has similarities to the visual photoreceptors (Fauré-Frémiet, 1958; Fauré-Frémiet and Rouiller,

1957). If we are permitted to extend this argument we can look upon the eyespot–flagellum system as a primitive "retinal" cell.

We have already shown that the velocity of swimming is proportional to a number of light quanta absorbed at the eyespot, and that the swimming motion is energetically controlled by light absorption at the eyespot. The shape of the intensity-dependence curve in Figure 2.10 showing a gradual rise with increasing intensity and the appearance of plateaus at higher intensity values is very similar to the current-intensity curve of a photoconductive cell. It has been suggested that the creation of nerve impulses in visual processes results from the production of electrical charge during the photoactivation of rhodopsin (Rosenberg, 1958). Whether a characteristic threshold potential really exists, and whether the energy transfer is done electrically, can be experimentally answered by measuring the potential drop between the eyespot and the flagellum. Such measurements, however, are extremely difficult to make.

In the photoreceptors of higher animals, the photoexcitation that triggers the optic nerve, and for the most part the energy contained in a nerve pulse, are both derived from chemical energy. Thus, the number of electronic charges involved in forming one such pulse is much larger than the minimum number of light quanta required to trigger the optic nerve. In the case of *Euglena,* however, such an amplification mechanism is not necessary. The energy carried in the minimum number of quanta required to excite the eyespot is comparable to the energy involved in the swimming motion. This means that one quantum of light absorbed at the eyespot can be associated with approximately one electronic charge formed at the base of the flagellum. At the saturation intensity of about 40 fc, the swimming velocity is about 0.018 cm/sec in a medium of viscosity of 0.987 cp. Using the cross-sectional area of the eyespot, the intensity of 40 fc at wavelength 465 nm (which is equivalent to about 2×10^{14} quanta/cm^2-sec), and the average radius of the *Euglena* cell, we have estimated the threshold potential to be of the order of 0.01 to 0.1 mV. This is small compared with the values found for nerves of higher animals, but within the order of magnitude found for certain insect photoreceptors (Naka, 1960).

It will be interesting to see as this discussion progresses if suitable analogies in structure, chemistry, and physiology can be drawn between the eyespot–flagellum photoreceptor—which enables the organism to respond to the direction, intensity, and wavelength of light—and the photoreceptor system of invertebrates that possess imaging eyes.

III. THE COMPOUND EYE

As we proceed phylogenetically from the protozoa with their eyespot and flagellum photoreceptor system to coelenterates, flatworms, roundworms, and segmented worms, we find a variety of photoreceptor structures ranging from eyespots to photosensory cells to simple "pin-hole" eyes. In arthropods and molluscs we find, in addition to these more primitive receptors, imaging compound eyes and refracting eyes. We can see that in the course of evolution invertebrates have used every known optical device for light detection and for forming an image (Figures 3.1–3.3).

Imaging Eyes

Of the various kinds of imaging eyes the pin-hole eye is the simplest. It uses the principle by which a small hole in the wall of an opaque chamber

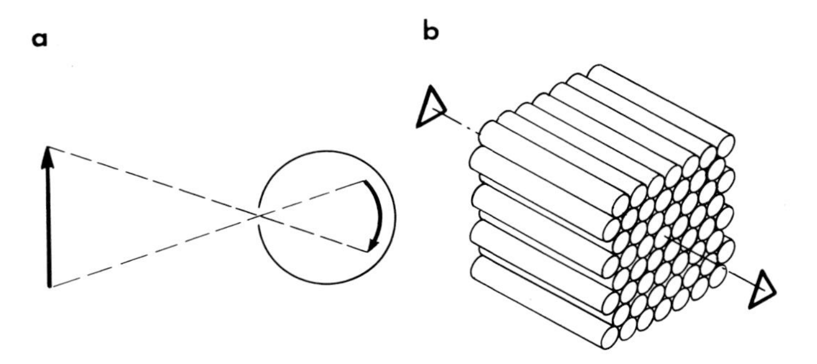

Figure 3.1. Imaging systems. (a) pin-hole; (b) parallel tubes.

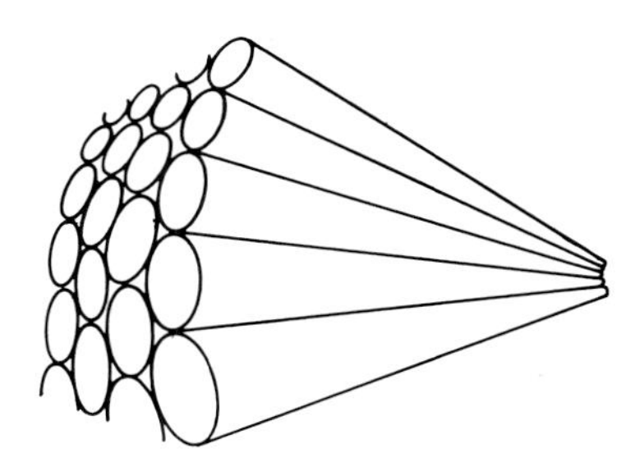

Figure 3.2. Arrangement of tubes as in the compound eye.

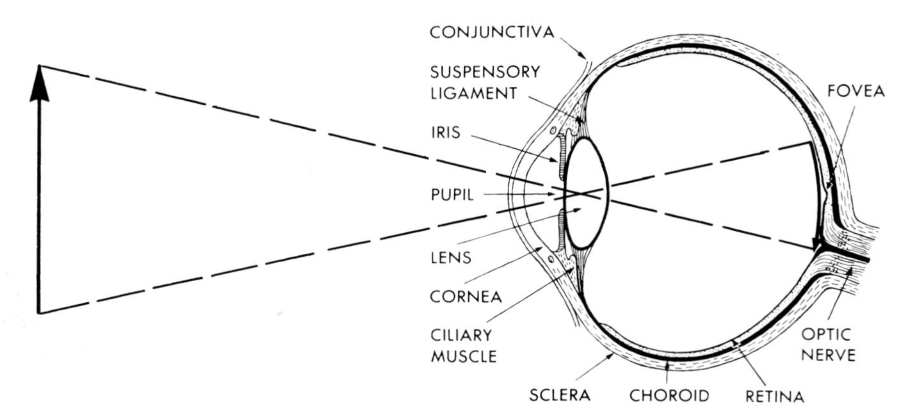

Figure 3.3. Human eye.

allows the passage of a very narrow beam of light from each point on an object, forming an inverted image on the opposite wall of the chamber (Figure 3.1a). As an image-forming device it is not very efficient, for only a small fraction of the light from an object can reach the photoreceptor surface. If the hole is made larger to increase the amount of light, image definition is lost, and if it is made smaller to improve the definition, diffraction effects become a problem. This kind of pin-hole eye is found in *Nautilis,* a cephalopod mollusc. It has the advantage of simplicity because no focusing is required for near or distant objects, and the size of the image is inversely proportional to the distance of the object.

A second kind of rudimentary optical system forms an image through a bundle of tubes which are separated by opaque walls. Only light falling on a particular tube in the direction of its axis can proceed to the end of the tube where it forms an upright image of the object (Figure 3.1b). The image formed is the same size as the object, regardless of distance.

Therefore, only those objects equal to or smaller than the device can be imaged completely, and no impression of distance by perspective can be conveyed since the image size remains unchanged with object distance. However, if the tubes are disposed around the surface of a sphere or sphere segment with their axes pointing toward the center, as is the case in compound eyes, many of these disadvantages are avoided (Figure 3.2).

For most vertebrates, including man, a third kind of optical system is used—a refracting eye (Figure 3.3). In this system the image is formed by the refraction of light at one or more spherical surfaces that separate media of different refractive indices. The refracting structures are the clear *cornea* and the *lens.* The refracting eye has the great advantage that image formation occurs through an integrative action, so that all rays falling on the eye from a given point of origin are brought to a point of focus on the retinal photoreceptors. The image produced there is inverted, and its size is inversely proportional to the distance of the object. Refracting eyes are also found in some invertebrates; for example, in mollusc cephalopods, the *Octopus* and the Squid.

In Annelids, for example, the earthworm, light-sensitive cells embedded in the body wall contain a lens surrounded by a neurofibrillar network. In many leeches such cells are gathered into cups shielded by pigment cells. These cup-shaped clusters of light-sensitive cells may lie under a lens, which is a specialized area of cuticle tissue. The complete structure of a lens and photosensory cell clusters is referred to as an *ocellus* (Figure 3.4).

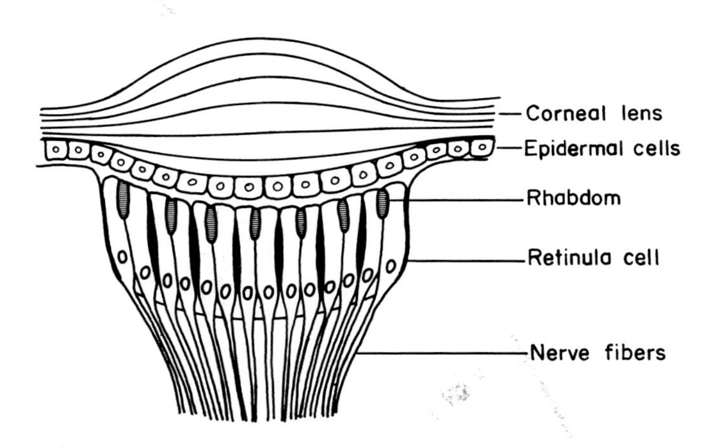

Figure 3.4. Schematized ocellus.

In Platyhelminthes (flatworms) such as planaria, we typically find two ocelli comprised of photosensory cells containing pigment granules (Figure 3.5). The pigment granules shade the sensory cells from light in all but one direction, and so enable the animal to respond differentially to the direction of light; that is, to turn from the light to the dark. The animal's tendency to avoid light seems to be controlled by the balance of nervous impulses from the eye. The photosensory cells have differentiated structures that resemble retinal rods. They are about 5 μ in diameter, with an average length of approximately 35 μ, and they consist of lamellae (Wolken, 1958a; 1961b). Similar photosensory cell structures have been found in the planaria *Dugesia lugubris* and *Dendrocoelum*

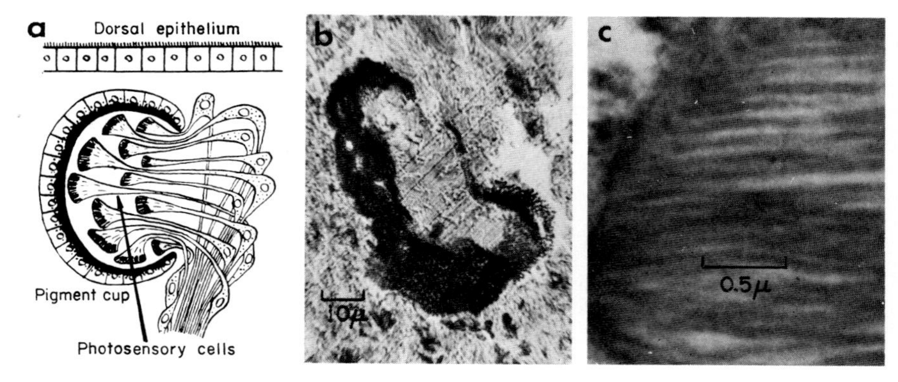

Figure 3.5. Planaria eye. (a) schematic of eye. (b) Section through pigment cup, light micrograph. (c) photosensory structure, electron micrograph.

lacteum (Röhlich and Török, 1961), and in the marine planarian *Convoluta roscoffensis,* peculiar to the coast of Brittany, France (Keeble, 1910).

The arthropods, which include insects, arachnids and crustacea, are the most numerous and diverse of the invertebrates. Though many possess eyespots, simple eyes, and ocelli, almost all arthropds have well-developed, image-forming compound eyes.

Compound Eye Structure

Compound eyes are good only for short range vision of about a few millimeters. But they are particularly efficient for detecting movements in their total visual field.

In the generalized structure of the compound eye (Figure 3.6) the eye facets are called *ommatidia.* The number of ommatidia varies from only a few in certain species of ants, to more than 2000 in the dragonfly. Each ommatidium is a complete eye in the sense that it has a corneal lens, crystalline cone, and retinula cells. Depending upon the insect, there are from three to eleven retinula cells per ommatidium. The retinula cell has a differentiated photoreceptor structure, the *rhabdomere,* which is analogous in function to the retinal rod outer segment of the vertebrate eye (Figures 5.3, 5.4, and 5.5). Collectively, the rhabdomeres form the light-sensitive *rhabdom,* which serves as the photoreceptor within each ommatidium.

Exner (1891) described two anatomically distinct types of compound

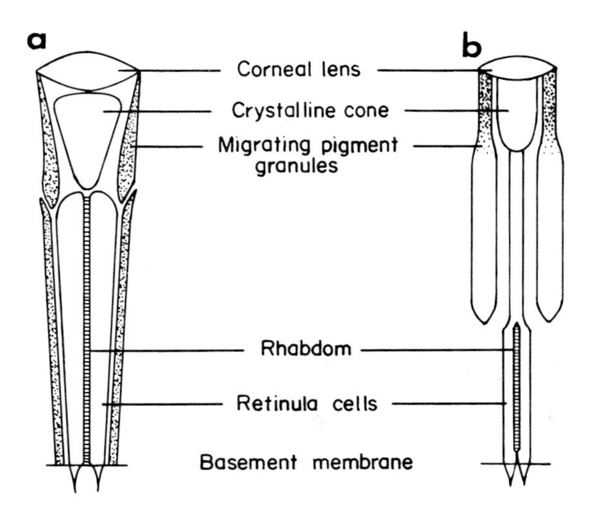

Figure 3.6. Compound eye ommatidium, showing the two anatomical types: (a) apposition; (b) superposition.

eye structures, the *apposition* and *superposition* eyes. In Exner's model, apposition eyes were those in which the rhabdomeres forming the rhabdom lay directly beneath or against the crystalline cone, and in which an inverted image was formed at the level of the rhabdom (Figure 3.6a). Each ommatidium was entirely sheathed by a double layer of pigment cells. Therefore, only light striking the lens within 10° of the perpendicular reached the rhabdomeres. Light striking the lens at a more oblique angle was either reflected by the lens or absorbed by the pigment sheath. Light falling upon the lens of an ommatidium reached only the rhabdomeres of that ommatidium. This type of compound eye structure was believed to be characteristic of diurnal insects.

Exner's concept of a superposition eye was one in which the receptor cells resided a certain distance away from the crystalline cone (Figure 3.6b). The term "superposition eye" has been loosely applied to any compound eye in which, during the dark-adapted state, the tips of the crystalline cones are separated from the rhabdom layer by an apparently clear space.

The superposition eye was believed to be characteristic of nocturnal species, and the superposition mechanism was thought to be important for the requisite increase in light-gathering power. However, superposition eyes have been found in both diurnal and nocturnal species. Thus, whether the superposition eye actually functions as Exner described has recently been questioned (Goldsmith, 1964; Horridge, 1969).

Insect Compound Eyes

Cajal (1918), famous for his work on tracing out the nervous systems of vertebrates, thought that the general plan for all visual systems would be found in the insect eye. After studying the insect eye, he wrote: "The complexity of the nerve structure for vision is even in the insect something incredibly stupendous. . . ."

Being fully forewarned, let us use the electron microscope to look at the compound eye of a few selected insects and see if we can find a structural basis for how they function for vision.

Drosophila melanogaster

The eye of *Drosophila melanogaster,* commonly referred to as the fruitfly, is composed of over 700 ommatidia (Figure 3.7). Each ommatidium consists of a corneal lens, a crystalline cone, retinula cells, and a sheath of pigment cells that extends over its entire length (Figure 3.8a). The ommatidium is approximately 17 μ in diameter and from 70 to 125 μ in length. In cross section there are seven retinula cells radially arranged. Each retinula cell has a medial portion extending toward the center of the ommatidium and terminating in a dense circular rhabdomere (Figure 3.8b and c). Each rhabdomere is distinct with respect to its retinula cell, and appears to be attached to it with a fine membrane. This

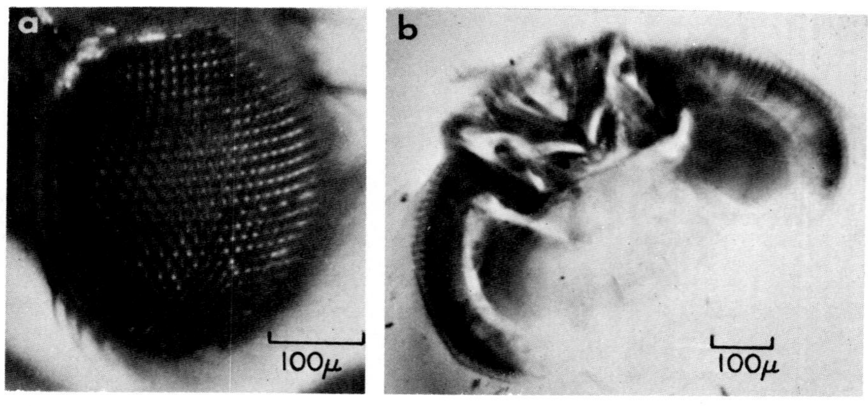

Figure 3.7. Compound eye, *Drosophila melanogaster.* (a) Photograph of the living eye; (b) a section through the whole head, showing both eyes. X-ray micrograph.

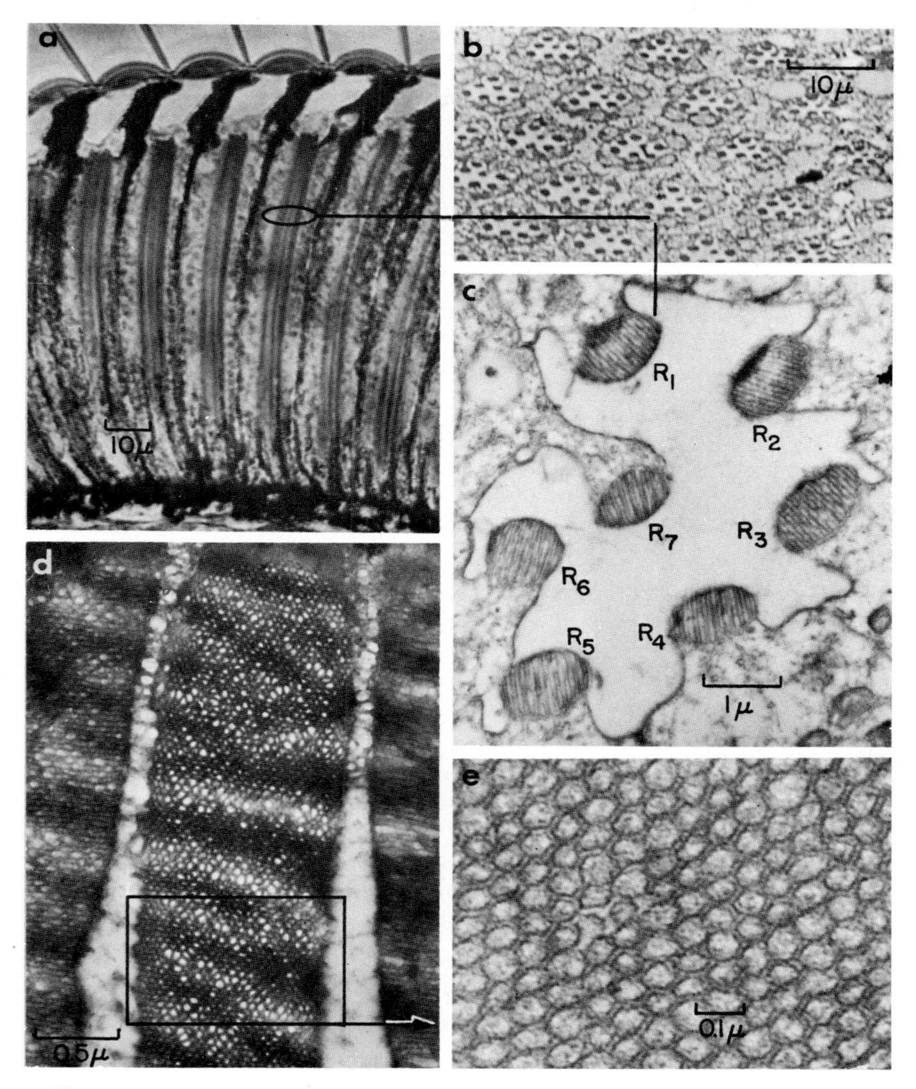

Figure 3.8. Compound eye, *Drosophila melanogaster,* light and electron micrographs. (a) Longitudinal section through several ommatidia showing the corneal lens, crystalline cone, rhabdom, and pigment sheath. (b) Cross section through ommatidia. (c) Cross section through the rhabdom to illustrate the orientation of the rhabdomeres and their fine structure. (d) Longitudinal section at the distal end of the ommatidium showing three adjacent rhabdomeres and their fine structure. (e) Enlarged area of a rhadomere to show the geometry of the microtubules.

structural relationship is strongly reminiscent of the attachment between the outer and inner segments of the vertebrate rod.

The rhabdom consists of seven individual rhabdomeres, R_1–R_7, situated in a relatively clear fluid cavity, one of which is asymmetrical. The rhabdomeres average 1.2 μ in diameter and are more than 60 μ in length. A definite fine structure of lamellae, separated by less dense interspaces, is found within each of the rhabdomeres. These dense laminations originate at the line of attachment and terminate in a scalloped border on the medial side of the rhabdomere. This lamellar structure is observed in all cross sections of the rhabdomeres, while the tubular structure is seen in some oblique and longitudinal sections (Figures 3.8d and e). A single structure of microtubules (Figure 3.8e) can produce both of these geometric structures in thin section, depending upon the orientation of the individual rhabdomere with respect to the plane of cutting (Wolken *et al.,* 1957a). This is schematically shown in Figure 3.9. Each of the rhabdomere microtubules is about 500 Å in overall diameter and has a double membrane, with a wall thickness of the order of 100 Å. A similar structural arrangement of seven rhabdomeres that form the rhabdom and a *fine structure* of microtubules for the rhabdomeres is also found for the housefly, *Musca domestica* (Fernández-Morán, 1956; Eichenbaum and Goldsmith, 1968).

The Cockroach

The cockroach, a nocturnal insect, is believed to be one of the more primitive of the unspecialized insects. Electrophysiological studies have indicated that the cockroach eye contains two different spectral receptor systems and the question arises whether there are any related structural differences (Walther and Dodt, 1957, 1959; Walther, 1958). Therefore, the structure of the compound eye of two large species of cockroaches, *Periplaneta americana* and *Blaberus giganteus,* was investigated by electron microscopy (Wolken and Gupta, 1961).

There are approximately 2000 ommatidia in each of these compound eyes and the ommatidia are separated from one another by a pigment sheath. A cross section through the *Periplaneta* eye shows many ommatidia surrounded by pigment granules (Figure 3.10a). Each ommatidium appears to be made up of seven retinula cells.

The number of retinula cells in the upper and lower halves of the cockroach compound eyes does not differ, but there are minor variations in the relative size of the cells, which is in agreement with previous studies of Walther (1958). Each retinula cell is from 7 to 9 μ in diameter with a

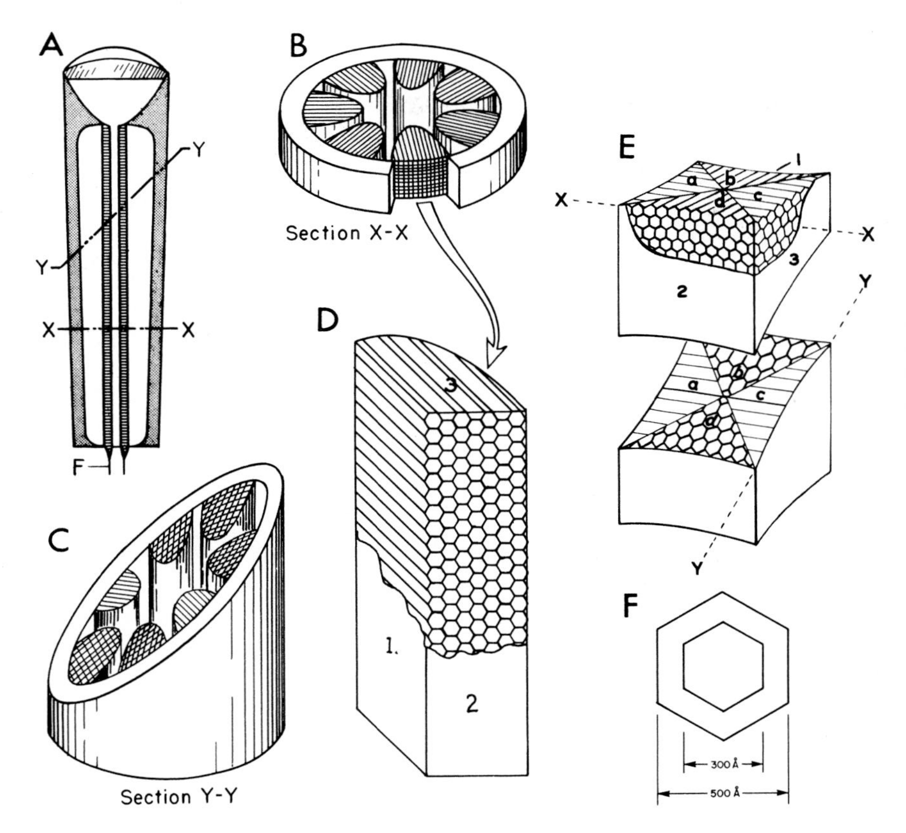

Figure 3.9. (A) Diagram of ommatidium. (B) and (C) are different sections of a rhabdom of the open-type; (D) structure of a rhabdomere; (E) rhabdom of four fused rhabdomeres a, b, c, d. (F) microtubule, average diameters.

large elliptical nucleus whose inner side is differentiated to form the rhabdomere. Although only seven retinula cells were observed in *Periplaneta,* eight retinula cells were found in the cockroaches *Blatta (Stylopyga) orientalis* and *Blatella germanica* (Jörschke, 1914; Nowikoff, 1932). The eighth cell is probably a rudimentary structure located close to the basal membrane (Dietrich, 1909). It does not extend the entire length of the rhabdom, and no rhabdomere is differentiated from it. Aggregates of intracellular pigment granules, which do not seem to be affected by dark-adaptation, surround the rhabdoms and extend the whole length of the retinula cells. Numerous mitochondria are dispersed throughout the retinula cell. Depending upon the location and angle of cut

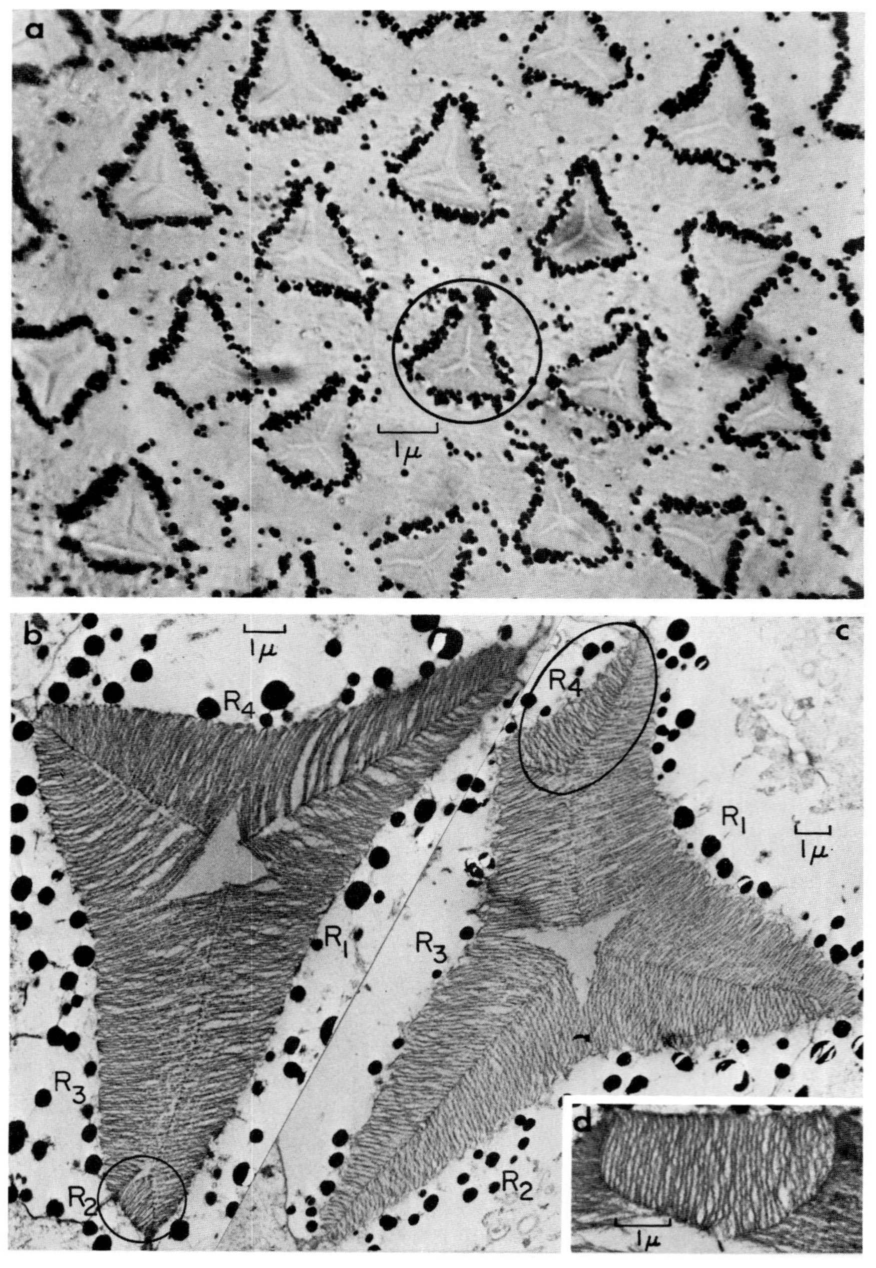

Figure 3.10. The cockroach, *Periplaneta americana*. (a) Light micrograph of a cross section through many ommatidia. (b) and (c) Cross section through two rhabdoms showing the arrangement of the four rhabdomeres (R_1–R_4) which form each rhabdom. (d) A cross section of one rhabdomere.

of the eye section, the rhabdom appears either rhomboid or triangular (with sides measuring from 5 to 12 μ in length). Each rhabdom is made up of four rhabdomeres (R_1–R_4) which lie in close proximity and form a regular pattern of organization (Figure 3.10b). This is schematized in Figures 3.9 and 3.11. A rhabdomere averages 2 μ in diameter and is about 100 μ in length. The cockroach rhabdomeres are also a single geometrical structure of packed microtubules. Each tubule is about 500 Å in diameter with walls of the order of 50 Å in thickness. There are approximately 400 tubules in one square micron of surface, or about 80,000 tubules in a single rhabdomere (Wolken and Gupta, 1961).

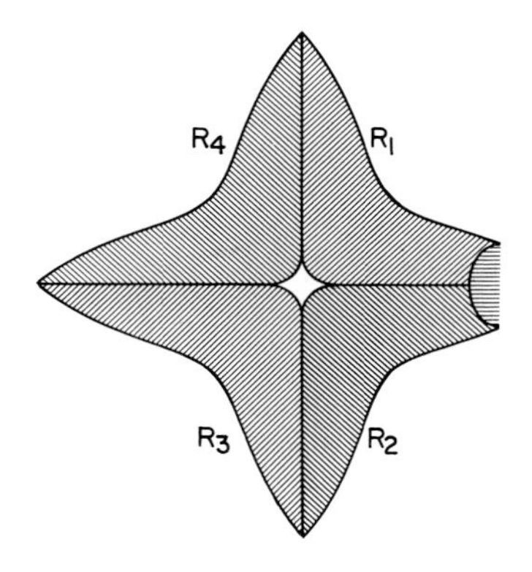

Figure 3.11. Diagram of a fused rhabdom similar to the cockroach, showing the four rhabdomeres (R_1–R_4).

The closed-type rhabdom observed here in the cockroach (Figure 3.10) is most likely an efficiency mechanism used for light capture by nocturnal insects. Evidence supporting this view is found in the fact that the cross-sectional area of the cockroach rhabdom is about five times that of the *Drosophila* rhabdom (Figure 3.8c).

The Firefly

The compound eye of the firefly provides an example of ommatidia which are structurally unique among insect eyes. The firefly was the model used by Exner (1891) to derive his theory of the superposition compound eye. However, the microanatomy of the firefly eye was never fully described, and only recently was the structure of the compound eye of *Photuris* studied by Horridge (1968, 1969).

All compound eyes are divided into two functional systems, the dioptric or optical system, and the photoreceptor or rhabdom system. In its optical system the firefly eye differs from other insect eyes in that the corneal lens extends into the region normally occupied by the crystalline cone. Eyes exhibiting this type of corneal lens structure have been called an exocone or pseudocone type of eye. We have investigated by electron microscopy the compound eye of *Photuris pennsylvanica,* which consists of several hundred ommatidia. The corneal lenses of the eye are seen in cross section to form spirals (Figure 3.12). In a longitudinal section these appear as laminations (Figure 3.13c). The corneal lens is surrounded by many distal pigment cells. Below the lens are situated the nuclei of four cone cells which extend to the basement membrane as fine filaments similar to the cone cell extensions found in other insects (Horridge, 1966; Perralet and Baumann, 1969; Waddington and Perry, 1960; Goldsmith, 1962). These filaments or cone cell threads (Horridge, 1968) are believed to function as wave guides directing light to the photoreceptor (Døving and Miller, 1969). They are seen as structures of increased electron density in an amorphous cellular medium where it is difficult to distinguish the boundaries of individual ommatidia (Figure 3.13d).

The receptor system of retinula cells that form the rhabdom occupies only one fourth of the ommatidial length. The system has two nuclear layers, a distal layer of retinula cell nuclei, and a basal layer of nuclei. For each ommatidium there appears to be a single basal cell which surrounds the axonal terminals of the retinula cells. These basal cells could serve to insulate the retinula cell axons in the manner of a Schwann cell. Each ommatidium contains eight retinula cells which form the double-layered rhabdom. The small distal rhabdom is formed by the rhabdomeres of two retinula cells. The major proximal portion of the rhabdom is formed from six retinula cells and has a six-arm configuration (Figures 3.14 and 3.15). Each arm is formed from the rhabdomeres of adjacent retinula cells. Each of the retinula cells of the proximal rhabdom has a V-shaped rhabdomere and little cytoplasm.

Firefly eyes are well adapted for light-gathering. The major portion of the rhabdom occupies nearly the entire cross-sectional area of the ommatidium and is tightly packed against the rhabdoms of neighboring ommatidia (Figure 3.14). In general, nocturnal insects such as moths and cockroaches (Fernández-Morán, 1958; Wolken and Gupta, 1961) have rhabdoms that occupy a large portion of the ommatidium, while daylight-active insects have relatively small rhabdoms (Wolken *et al.,* 1957a,b; Goldsmith, 1962). The almost solid layer of rhabdom found in fireflies, however, is as unique as the exocone and crystalline thread dioptric sys-

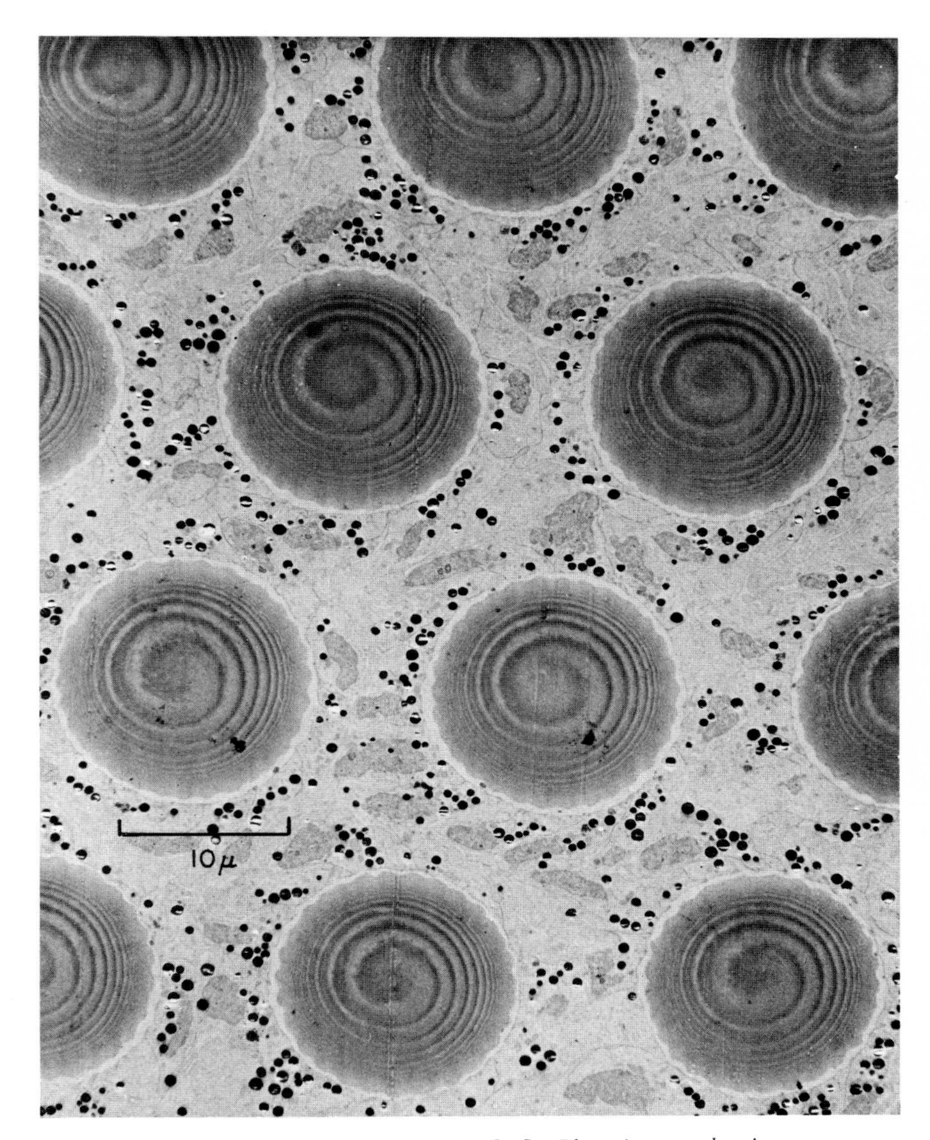

Figure 3.12. Lens structure of the firefly, *Photuris pennsylvanica.*

tem (Horridge, 1968, 1969). The ommatidium of the firefly is schematized in Figure 3.13. Compare it with the ommatidium of *Drosophila,* Figures 3.8 and 3.9.

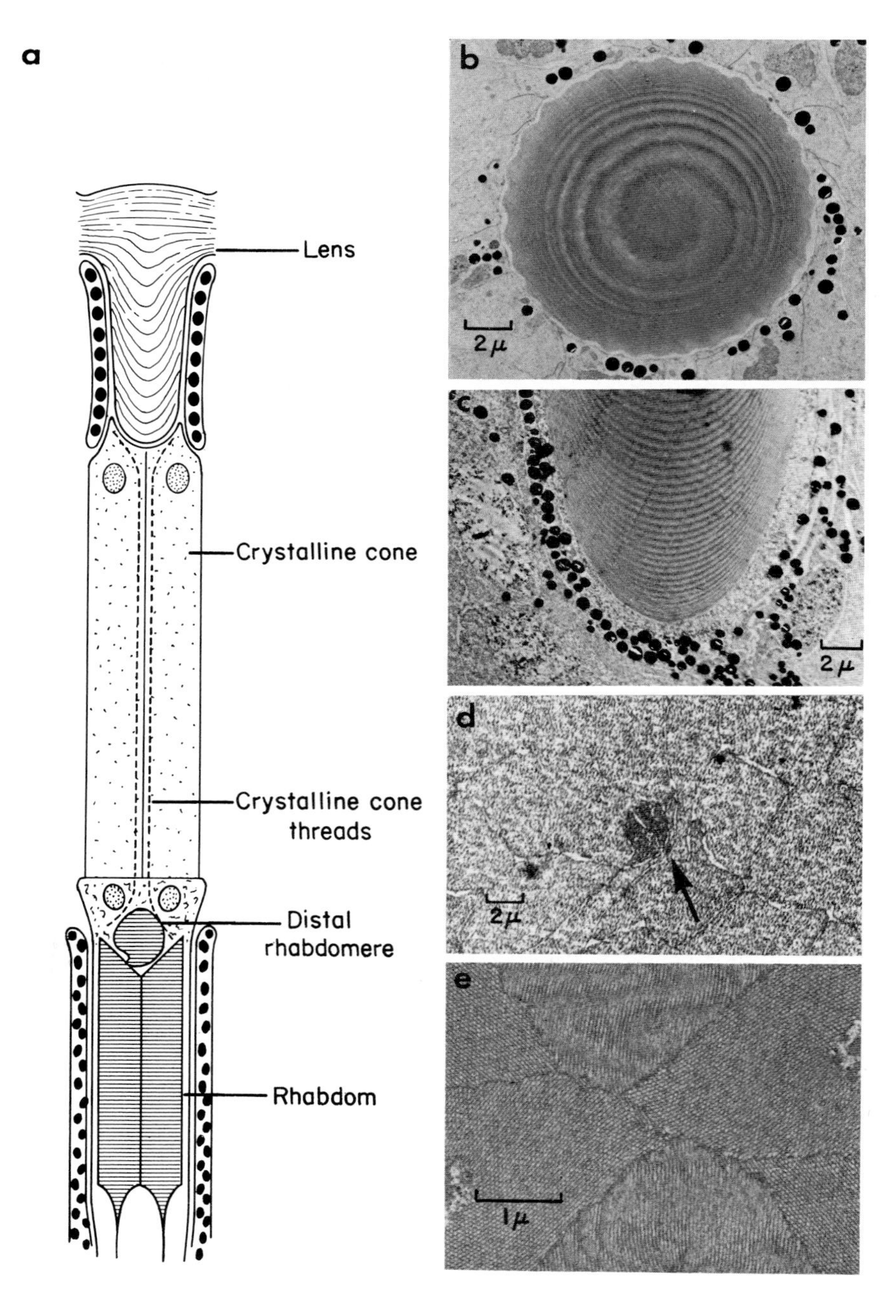

Figure 3.13. (a) Schematic structure of the firefly ommatidium. (b) and (c) Exocone lens. (d) Crystalline cone thread. (e) Rhabdom cross section, *Photuris pennsylvanica*.

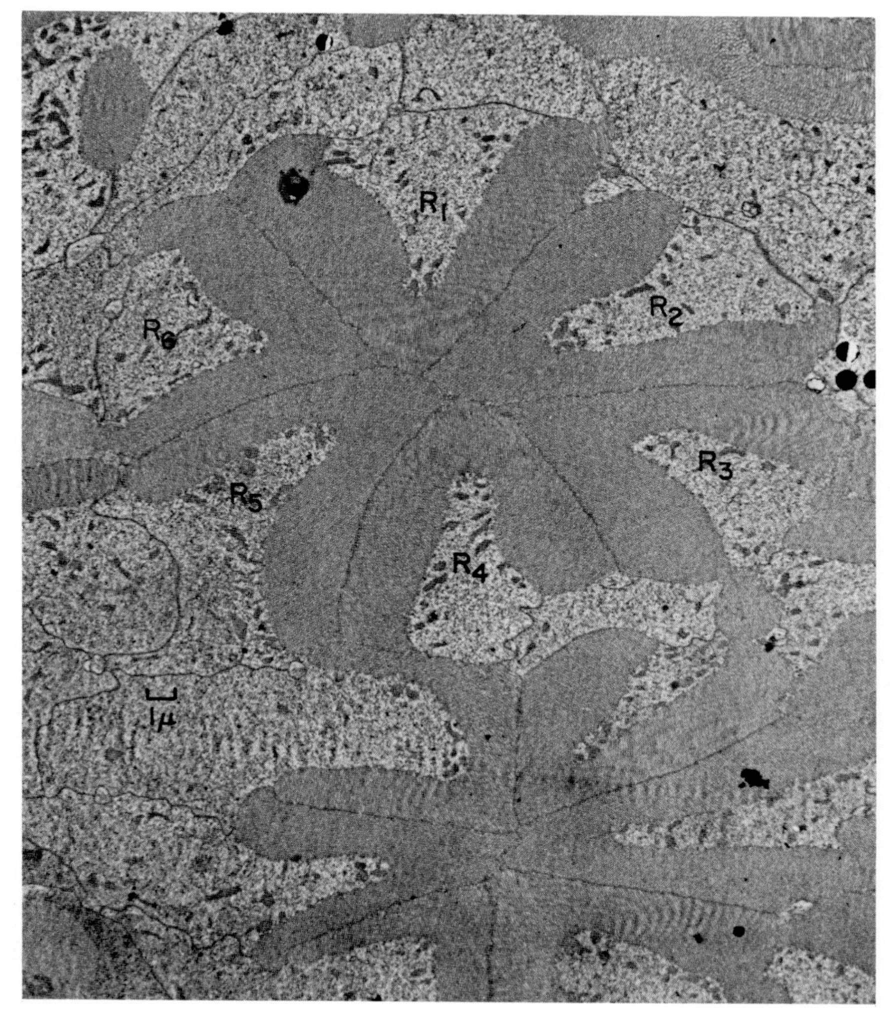

Figure 3.14. The firefly *(Photuris pennsylvanica)* rhabdom showing the geometrical arrangement of its rhabdomeres (R_1–R_6).

Other Insect Photoreceptors

An ommatidium from the compound eye of the baldface hornet, *Vespa maculata,* contains eight retinula cells (Figure 3.16a). In a few ommatidia there is a ninth retinula cell located more distal than the other eight. The rhabdom here is composed of four pairs of rhabdomeres (Figure 3.16b).

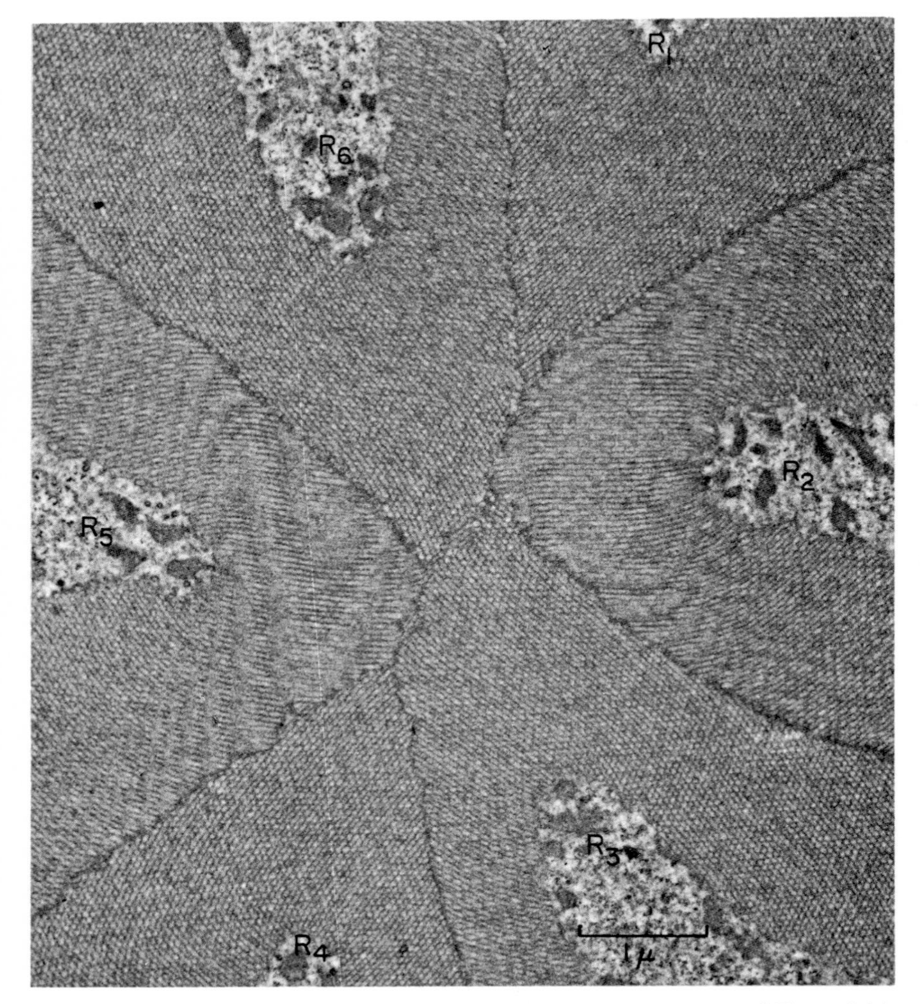

Figure 3.15. Rhabdom of firefly *Photuris pennsylvanica*. Enlargement of Figure 3.14, showing fine structure of rhabdomeres.

The crystalline cone is built of four cellular segments which taper into long tubular cone cell extensions (this is also found for the clothes moth, Figure 3.17) ending just distal to the basement membrane. There are no connections between these cone cell extensions and the nerve cells. On the other hand, the cone cell extensions have a definite structural relationship to the pigment cells which appear to make synapse-like connections with them.

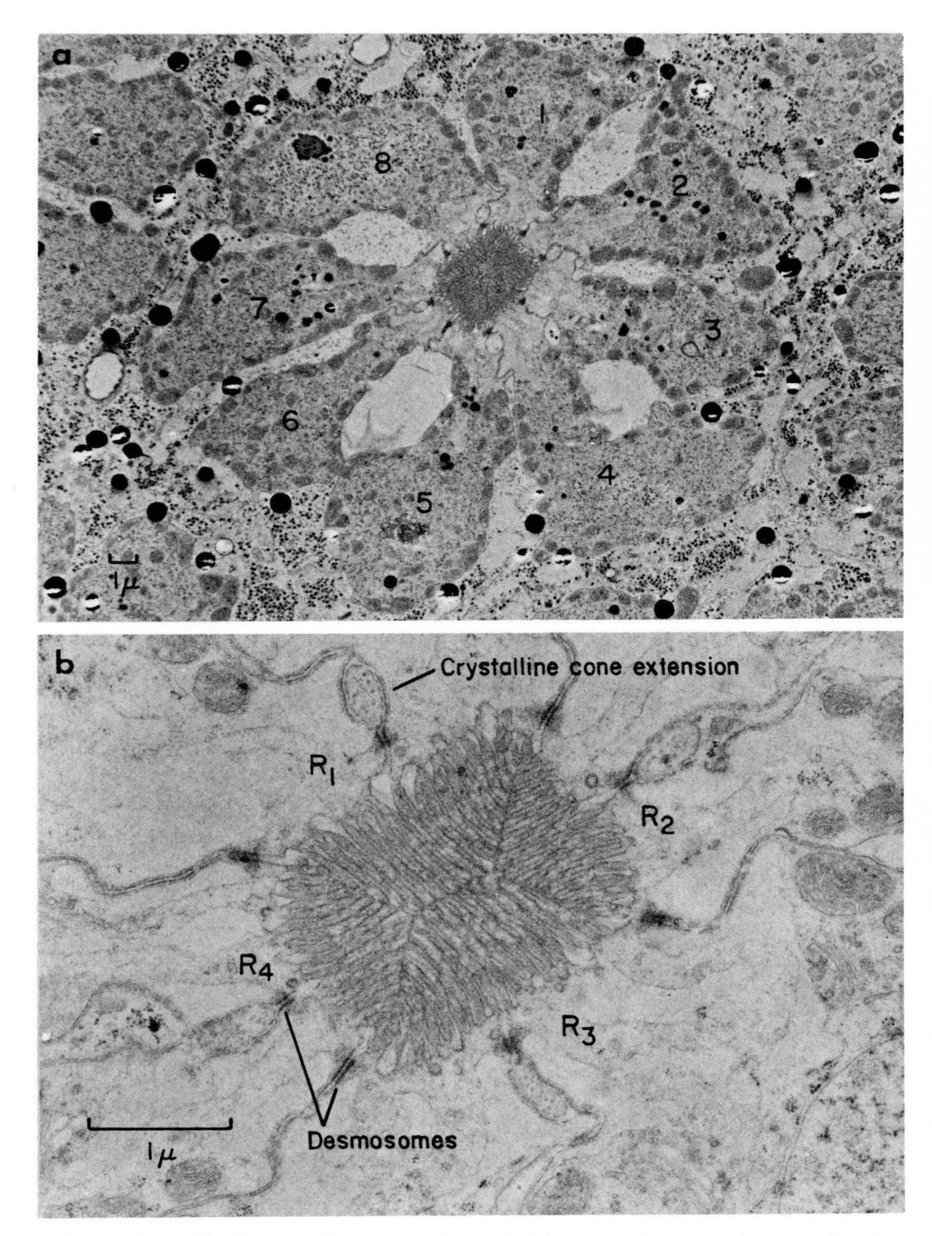

Figure 3.16. The hornet *(Vespa maculata)* rhabdom. (a) Cross section showing the retinula cells (1–8). (b) The central part of (a) in greater detail showing the fused rhabdom, formed by the rhabdomeres (R_1–R_4).

The compound eye structure of the honeybee, *Apis mellifera,* was studied by Goldsmith (1962). The eye of this worker bee has several thousand ommatidia. Cross sections of the rhabdom show it to consist of four rhabdomeres, although it is formed from eight retinula cells. The rhabdom structure is similar to that of the hornet (Figure 3.16b) just described. It should be noted that each rhabdomere is a closely packed parallel array of microtubules with their axes perpendicular to the axis of the rhabdom. The microtubules in adjacent rhabdomeres of the rhabdom are mutually perpendicular.

In the carpenter ant, *Camponatus herculenus pennsylvanicus,* the compound eye structure appears to represent a transition from the apposition-type eye (Figure 3.6a) to that of the pseudocone described for the firefly

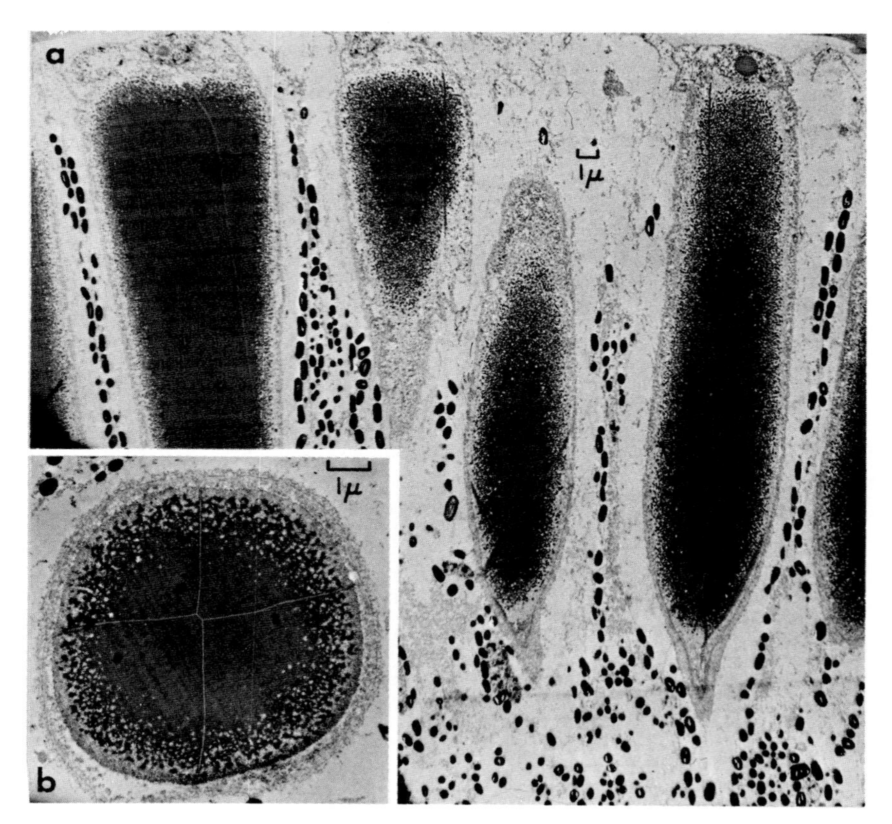

Figure 3.17. The clothes moth *(Tineola biselliella)* crystalline cone. (a) Longitudinal section. (b) Cross section.

(Horridge, 1969), and the crystalline thread of the hornet (Figures 3.13 and 3.16).

Like the hornet, the retinula cells of the carpenter ant show two zones, a clear zone near the rhabdom and a peripheral cytoplasmic zone. The rhabdom (Figures 3.18 and 3.19) of the carpenter ant is circular and occupies a much larger portion of the ommatidial cross section than in the hornet (Figure 3.16). Despite its size, the rhabdom is formed from only six retinula cells as compared with eight or nine in other insects.

The microtubules of the rhabdomere have an elaborate fine structure. There appears to be a lobulated intratubular material at the center of which is a star-shaped electron dense body (Figure 3.18b). At the angles of adjacent microtubules, electron dense intervillous bridges are found. These intervillous bridges bind adjacent microtubules (Figure 3.19).

The nerve bundle from an ommatidium contains seven axons. Yet, we have only been able to identify six retinula cells in our sections. It is not clear whether the additional axon is from a vestigial retinula cell which our sections have missed, or whether this is a second order neuron. Vowles (1954) has described seven retinula cells in other species of ants.

Related Discussion

Although we do not completely understand how the compound eye functions in insect vision, our picture of the anatomical structure of the optical and photoreceptor systems is finally coming into focus.

From all the available structural information we can abstract three common features of all compound eyes which help us relate their structure to their function. We find that the rhabdom is either of the open type (Figure 3.8) or the closed type (Figures 3.10, 3.14, 3.16, and 3.18); that each pair of adjacent rhabdomeres has its microtubules aligned in perpendicular directions, but that opposite rhabdomeres have parallel microtubules; and that in all cases the rhabdomeres are composed of microtubules ranging from 400 to 500 Å in diameter.

Exner's idea of the superposition type compound eye which has been with us since 1891 is now beginning to appear very doubtful (Goldsmith, 1964; Horridge, 1969). Horridge prefers that compound eyes be classified anatomically in terms of *with or without crystalline threads* and *fused* or *separate rhabdomeres*. Post and Goldsmith (1965) preferred that superposition and apposition eye be replaced by scotopic and photopic, in order to more accurately describe the physiological properties of these photoreceptors.

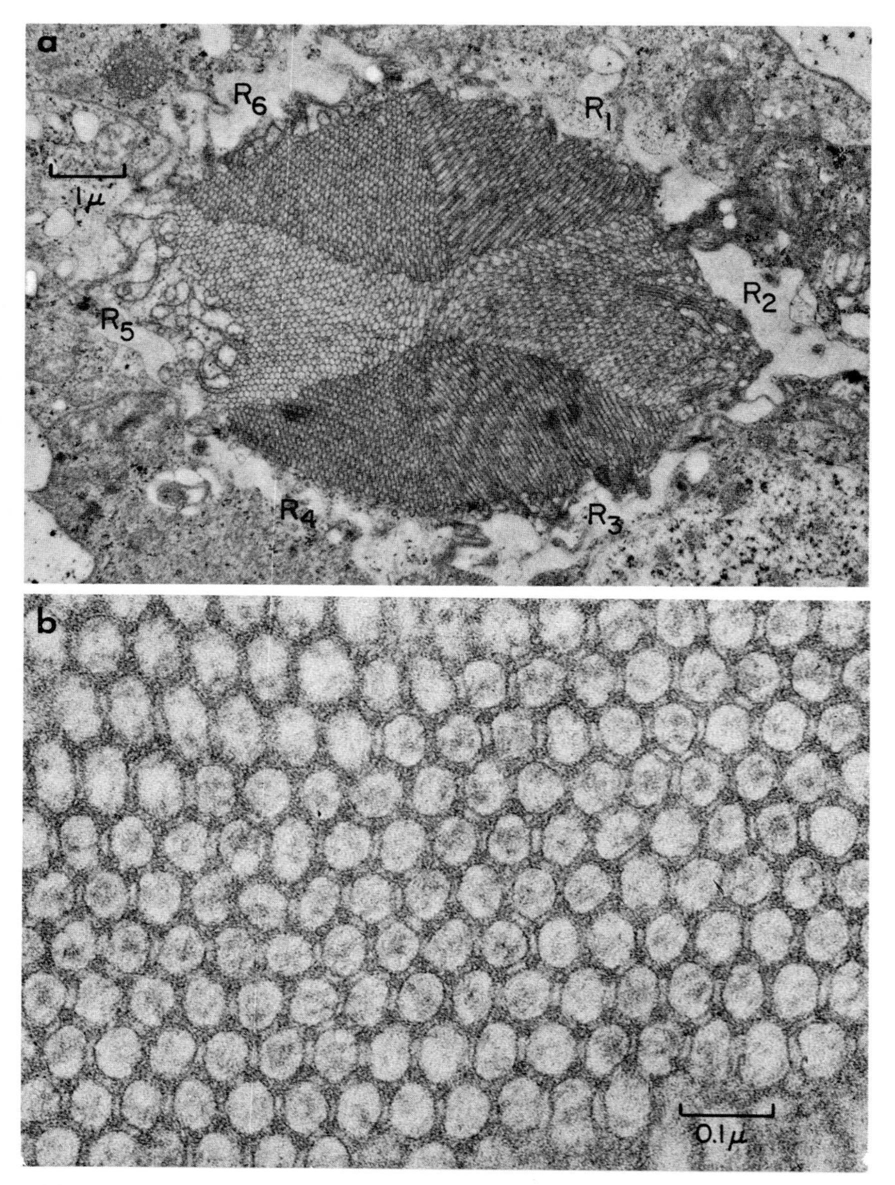

Figure 3.18. The carpenter ant *(Camponotus herculenus pennsylvanicus)*. (a) Cross section through the rhabdom showing its rhabdomeres (R_1–R_6). (b) High resolution electron micrograph of a cross section through many rhabdomere microtubules.

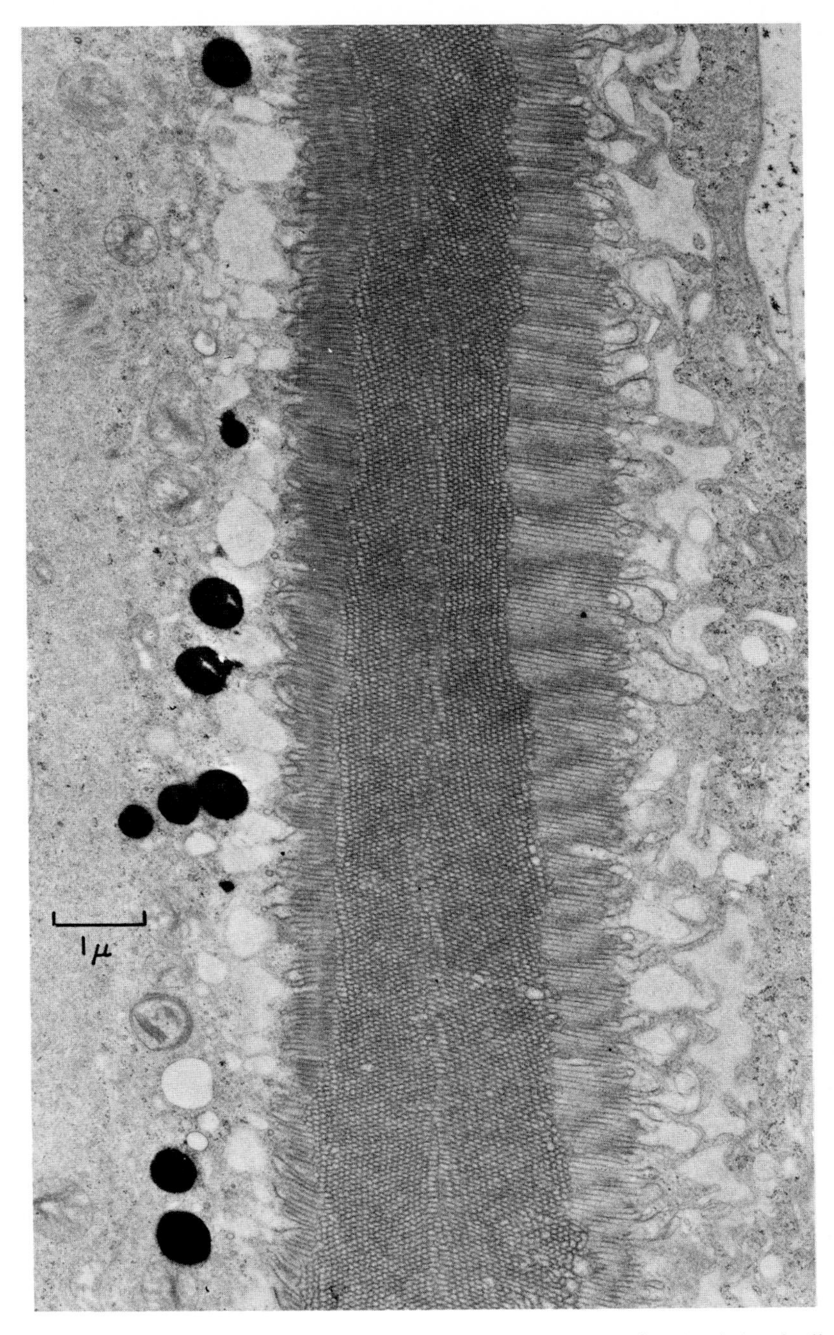

Figure 3.19. The carpenter ant *(Camponotus herculenus pennsylvanicus)*, longitudinal section of the rhabdom.

A most interesting observation is that in certain insect ommatidia, e.g., the firefly, there exist crystalline cone threads that could function as waveguides (Horridge, 1968, 1969; Døving and Miller, 1969). The evidence for this is that only the light contained within the waveguide (tract) is effective for stimulating the photoreceptors. If so, this would be inconsistent with the superposition model proposed by Exner (1891).

Another series of interesting studies that bear on the anatomical studies of the insect compound eye were those by Burtt and Catton (1962) and Rogers (1962). They performed experiments to determine the limit of resolution of the insect eye by using striped patterns and measuring the ERG (electroretinogram). They found a higher degree of resolution than had been predicted from behavioral studies. Their results implied a resolving power beyond that possible with an aperture the size of a single ommatidium. To account for this high resolution, Rogers (1962) proposed a theory of insect vision that involved complex diffraction images formed by the interaction of light from several ommatidia. According to this theory, the insect eye acts as a diffraction grating and produces several orders of images. In certain cases—depending on the length of the rhabdomere—it is possible for three of these images to lie within the receptor regions of the ommatidium. This process would involve the optical interaction of several ommatidia which would in turn increase the effective aperture and produce images which would allow the fine resolution that has been observed.

To resolve objects with an angular separation of 0.1 degree, the retinula cells of an ommatidium would have to function independently.

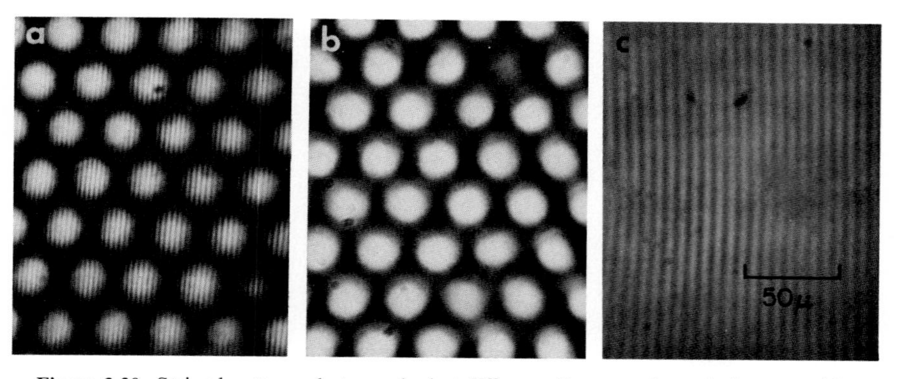

Figure 3.20. Striped pattern photographed at different distances through the corneal lens and crystalline cone of the grasshopper. (a) First-order image; (b) second-order image; and (c) third-order image.

The optical images of a striped pattern formed at various distances by the corneal lenses and crystalline cones of a grasshopper eye were photographed through a microscope and are presented in Figure 3.20. The formation of these images depends on light passing between ommatidia so that the fields of adjacent ommatidia overlap. The extent to which light can escape from one ommatidium into its neighbor in the compound eye has been examined by electrical recordings from single receptors during stimulation of single ommatidia by Shaw (1969). In the apposition eye of the drone honeybee and locust, interaction is extremely small. In the superposition type eye more than half of the light captured by the average retinula cell enters from neighboring ommatidia, even when the screening pigments are in their fully light-adapted position.

As a light moves across the visual field, at any given instant some of the individual retinula cells in an ommatidium "see" the light and some do not, while simultaneously some retinula cells of adjacent ommatidia also "see" the light and some do not (Kuiper, 1962). One of the difficulties with this hypothesis is that it cannot adequately account for the presence of screening pigment granules; such screening would prevent the requisite passage of light between ommatidia.

Further discussion must await an examination of the photoreceptor structures of crustacean eyes and knowledge of the nature of the visual and screening pigments of arthropods. Let us turn then directly to the eyes and photoreceptors of crustacea and molluscs.

IV. CRUSTACEA AND MOLLUSC EYES

Crustacea

A variety of animal forms are found in the Crustacea, from the lobsters and crabs which are the giants among them, to the water flea, only a few millimeters long. They possess simple eyes and compound eyes that vary in structure. Here, too, as in our discussion of the compound eyes of insects, we will describe several species of freshwater and marine crustacea that we have studied in our laboratory.

Daphnia

The freshwater *Daphnia (Daphnia pulex* and *Daphnia magna),* the water flea, possess two eyes, a compound eye and a simple nauplius eye (Figure 4.1). The compound eye was believed to be a relatively uncom-

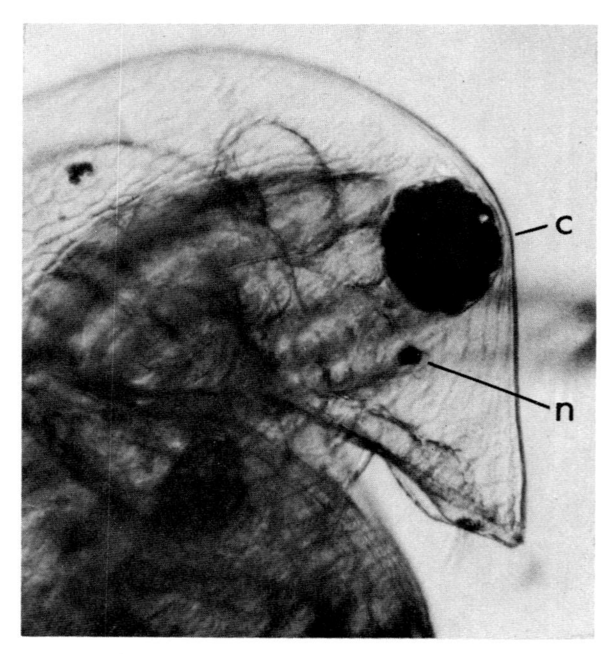

Figure 4.1. *Daphnia pulex.* The compound eye (c) and naupluis eye (n).

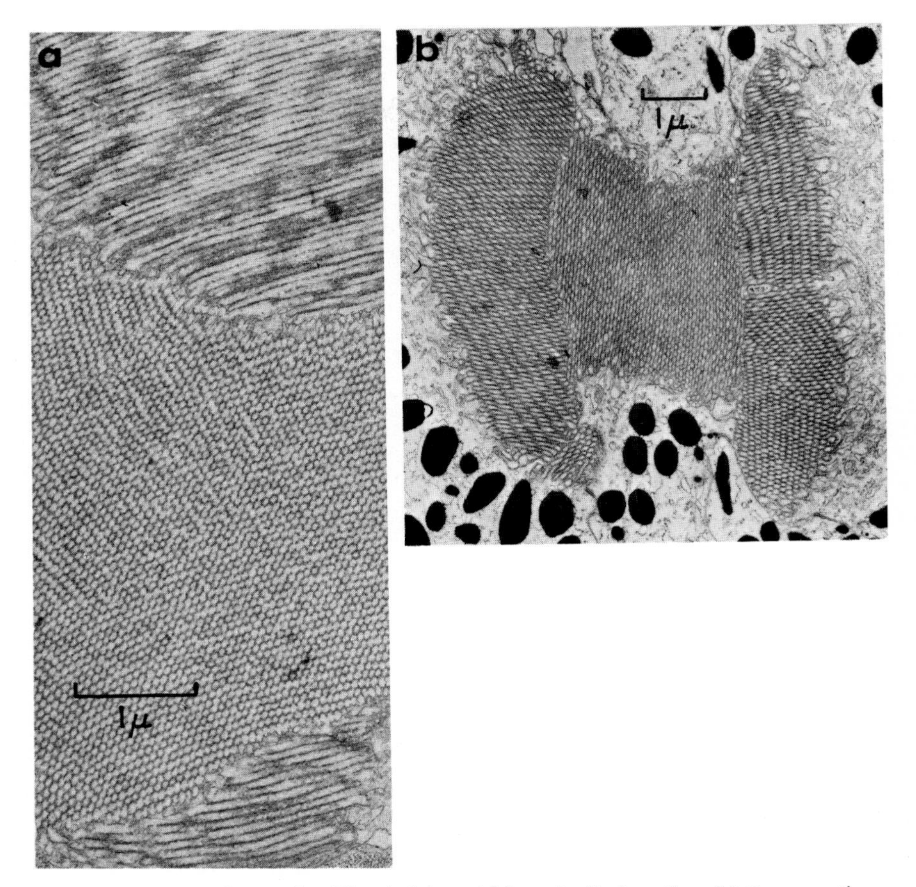

Figure 4.2. *Daphnia pulex.* The rhabdom, (a) Longitudinal section. (b) Cross section.

plicated structure, consisting of about 22 ommatidia enclosed in a capsule (Röhlich and Törö, 1965). The corneal lens, as seen in the insect ommatidium, and the distal pigment cells are not usually found (Figure 3.8). The ommatidium has a crystalline cone and elongated retinula cells, which are surrounded by pigment granules. There are eight retinula cells that give rise to seven rhabdomeres to form the closed-type rhabdom.

The rhabdomeres that form the rhabdom consist of microtubules or lamellae about 500 Å in diameter, and double-walled membranes about 100 Å thick (Figures 4.2, 4.3, and 4.4), not unlike what is observed for the insect. The microtubules within each rhabdomere are precisely arranged with their longitudinal axes regularly aligned in a given direction for one set of rhabdomeres, and in the perpendicular direction for alternate

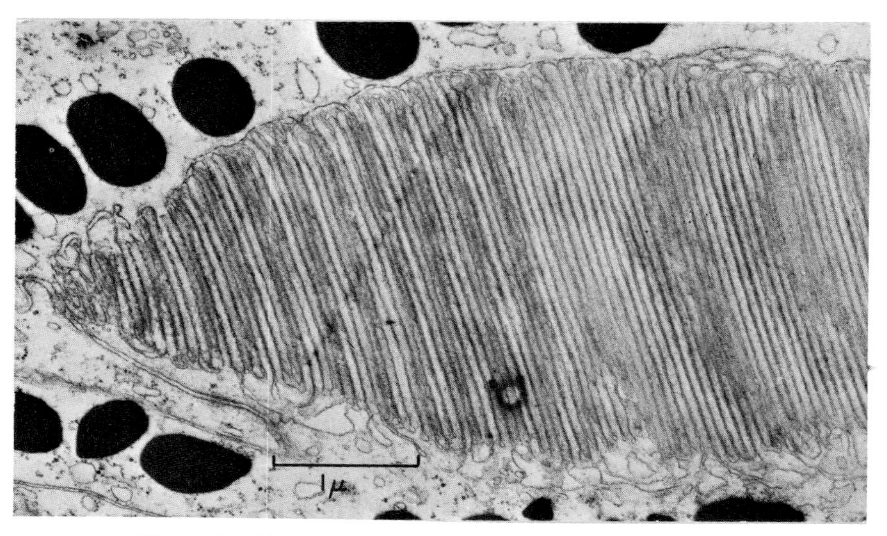

Figure 4.3. Rhabdomere, longitudinal section. *Daphnia pulex.*

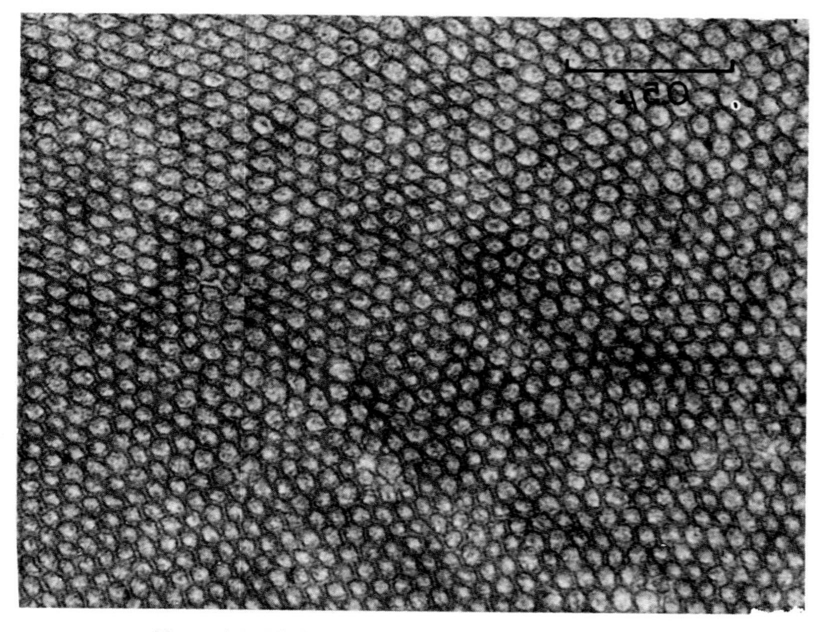

Figure 4.4. Rhabdomere microtubules. *Daphnia pulex.*

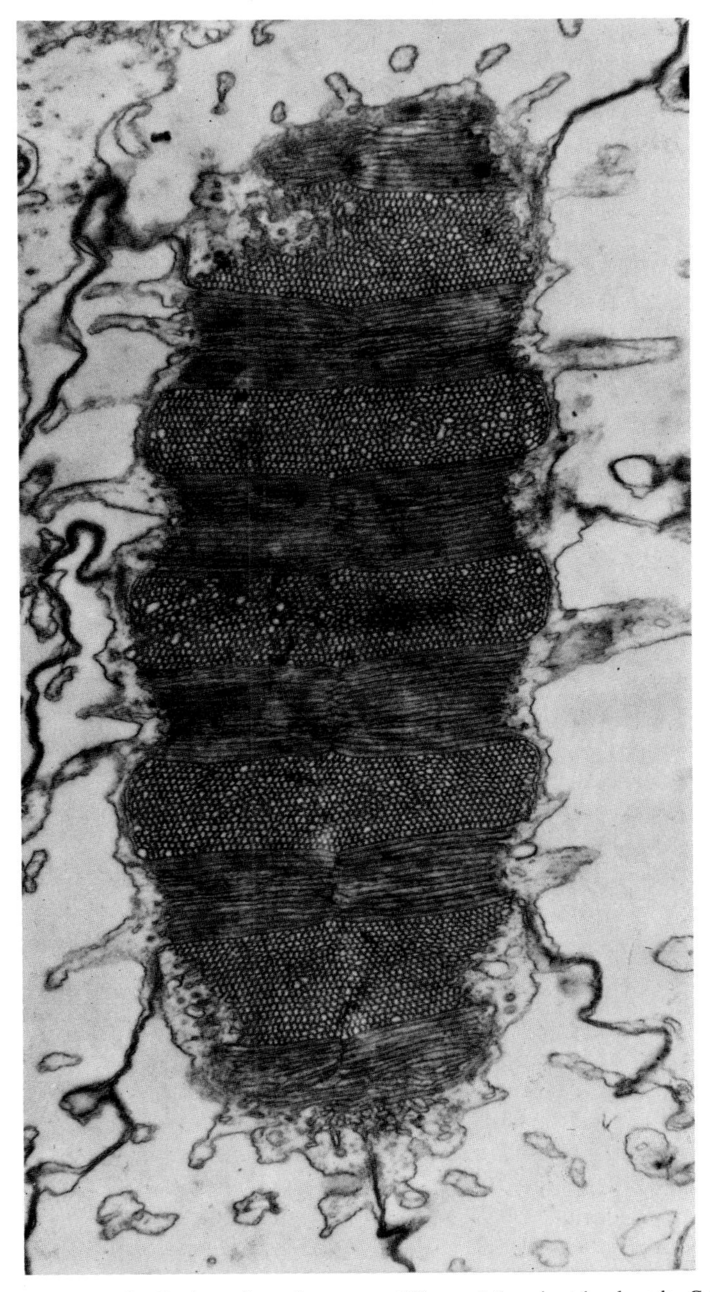

Figure 4.5a. Longitudinal section of an ommatidium of the giant land crab, *Cardisoma guanhumi.* Note the alternating layers of microtubules and lamellae oriented at 90° in successive layers (from Waterman, 1966, p. 33).

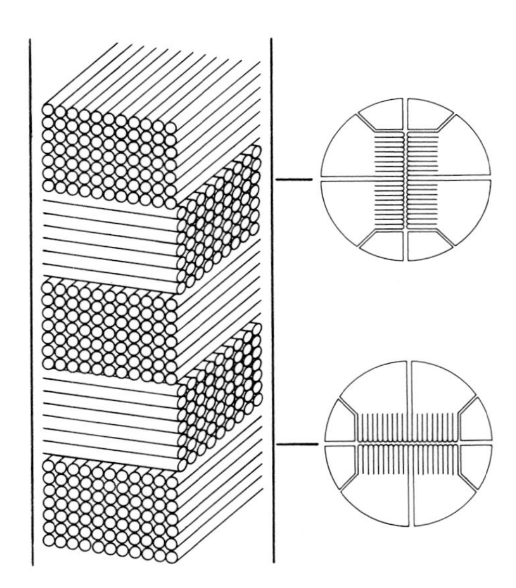

Figure 4.5b. Reconstructed rhabdom, showing orientation of microtubules. Cross section of rhabdom at two different levels, as found in *Daphnia* and some other crustacea (from Waterman, 1966, p. 35).

layers (Figure 4.2a). This type of structure was observed for other crustacea, for example, in the land crab, *Cardisoma* (Figure 4.5a) and the swimming crab, *Callinectes,* and is schematized in Figure 4.5b (see Eguchi and Waterman, 1966; Waterman, 1966). Such an arrangement of the rhabdomeres with their microtubules at right angles could be a structural basis for the ability of *Daphnia* and other crustacea to analyse the direction of polarized light (Baylor and Smith, 1953; Waterman *et al.,* 1969).

Leptodora

Turning to another freshwater crustacean, the carnivorous *Leptodora kindtii* is relatively large (18 mm in length) in comparison with *Daphnia*. It is almost completely transparent with one median spherical compound eye (Figure 4.6) with various types of connecting neurons (Gerschler, 1911, 1912; Scharrer, 1964). *Leptodora* is believed to have superior vision to most crustacea. The basis for this is the rapidity with which it captures for food copepods as large and fast as *Cyclops*.

The entire eye of *Leptodora kindtii* is contained within the transparent exoskeleton at the anterior end of the organism (Figure 4.6). The eye is free to move and can rotate 10° in either direction. A small area in back of

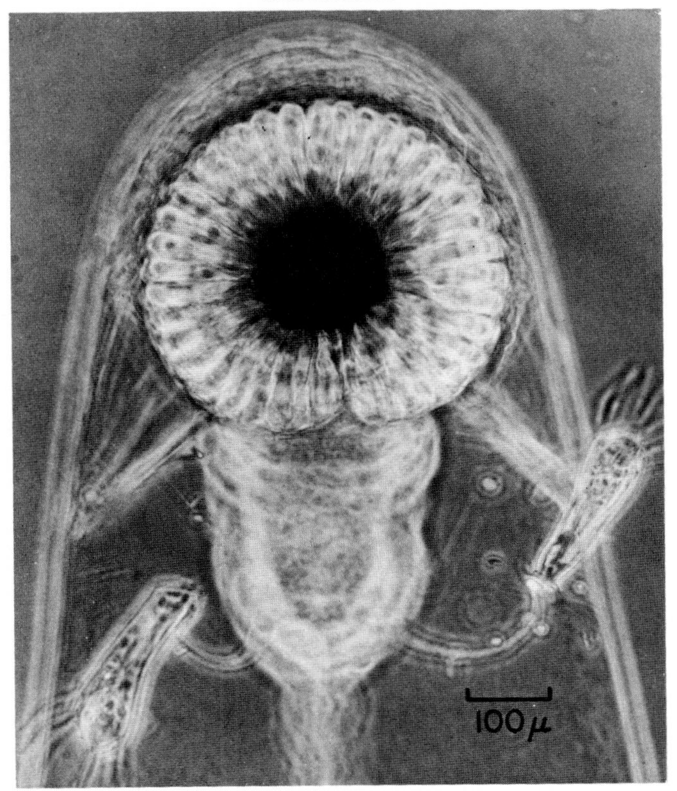

Figure 4.6. The eye of *Leptodora kindtii*.

the eye is for accommodation of the optic processes, and the brain lies in close proximity to these eye structures.

The *Leptodora* compound eye is composed of approximately 500 ommatidia radially arranged (Figure 4.6). The ommatidia are large conical structures, 180 μ in length and from 30 μ in diameter at the outer portion to about 2 μ at the base. A schematic ommatidium is illustrated in Figure 4.7a, and cross sections through the crystalline cone and rhabdom are seen in Figures 4.7b,c, and d. The crystalline cone constitutes about two-thirds of the ommatidial length. Although it is rounded at the outer end, there is no evidence of a distinguishable lens cap. There is, however, a completely transparent interstitial space between the surface of the eye sphere and the external chitinous wall, which probably functions as a common lens for all the ommatidia. The crystalline cone is composed of five equal pie-shaped segments, formed from five crystalline cone cells which serve to concentrate the light into a narrow beam (Figure 4.7b).

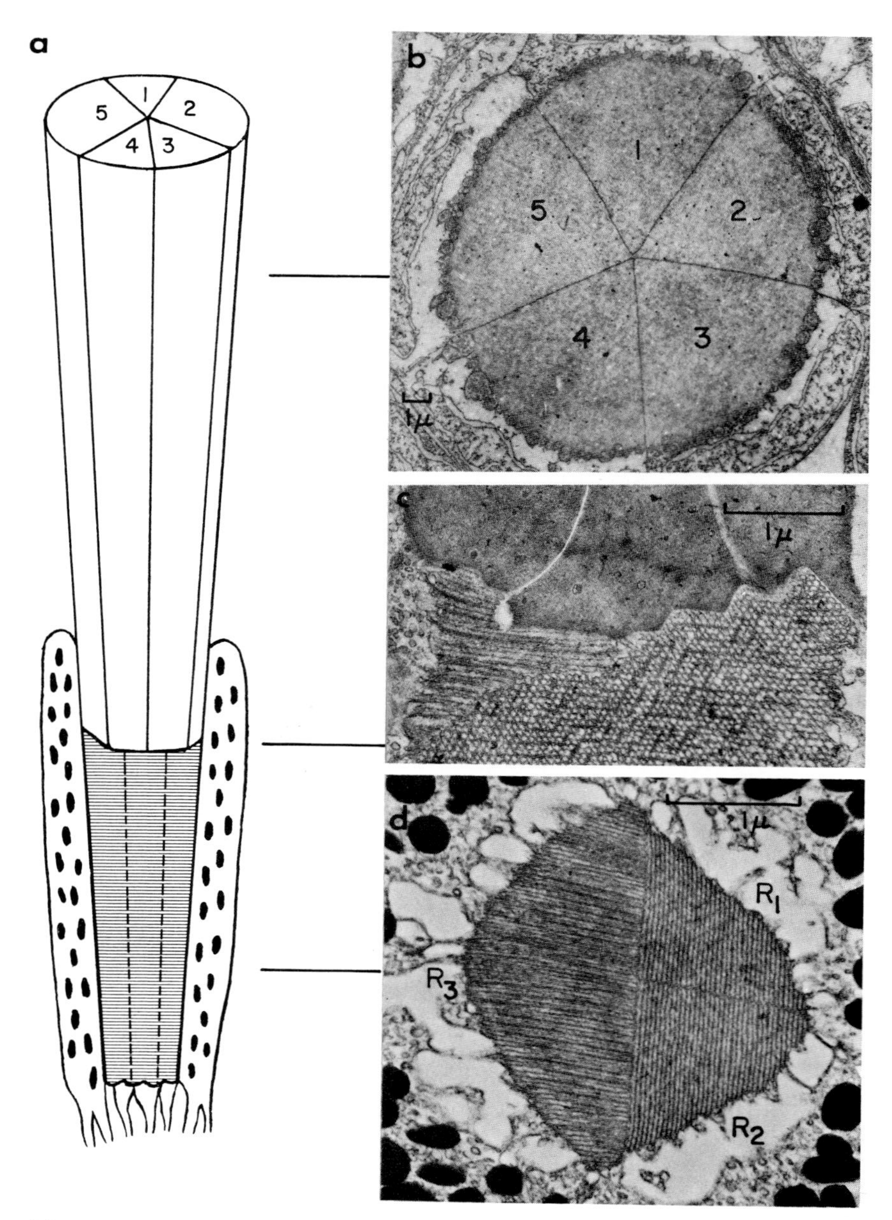

Figure 4.7. (a) *Leptodora kindtii* ommatidium. (b) Crystalline cone. (c) Longitudinal section of the connection of the crystalline cone with the rhabdom. (d) Cross section of the fused rhabdom showing rhabdomeres R_1–R_3.

Whether or not this segmentation promotes a system of total internal reflection is not known. Although structural data show that the crystalline cone continues proximal to the surface of the rhabdom as in the apposition-type eye, observations of pigment migration indicate that under certain conditions of dark-adaptation, "crossing" among adjacent crystalline cones could result in the formation of a superposition image. As the cones continue inward, the space between them increases and is filled with pigment cells.

The rhabdoms are affixed directly to the ends of the crystalline cones (Figure 4.7c). The four radially arranged retinula cells that form the rhabdom show only three rhabdomeres (Figure 4.7d). Cross sections through numerous ommatidia in the rhabdom area (Figure 4.8) reveal the rhabdom structure. One of the rhabdomeres (R_3) is large in comparison with the other two (R_1 and R_2) and appears to be two rhabdomeres that have fused (Wolken and Gallik, 1965). The fact that four retinula cells yield only three closed-type rhabdomeres for the rhabdom has also been observed for the dragonfly (Goldsmith and Philpott, 1957; Naka, 1960).

The rhabdomere fine structure of *Leptodora* is that of tightly packed microtubules (Figure 4.9), which average about 500 Å in outside diameter and whose double-membraned wall is about 150 Å wide. The microtubules of the small rhabdomeres R_1 and R_2 are arranged perpendicular to those of the large rhabdomere R_3. The ends of the microtubules appear continuous with the cytoplasm (Figure 4.10). This has also been observed for the rhabdomeres of many other arthropods, and the importance of this structural connection with the cytoplasm for visual excitation is discussed by Lasansky (1967).

Copilia

Another extremely interesting crustacean is the relatively rare copepod, *Copilia (Copilia quadrata,* Mediterranean; *Copilia mirabilis,* Caribbean). Its eye has been of considerable curiosity since it was first found in Naples and described by Grenacher (1879) and later, Exner (1891).

This copepod (which is about 1 mm wide and 3 mm long) is found in both the Mediterranean (Figure 4.11a) and Caribbean Seas (Figure 4.11b). Only the female of the species possesses the remarkable scanning eyes which make up more than half of its transparent body (Figures 4.11 and 4.12). Recently Vaissière (1961a,b), Gregory (1966, 1967) and Gregory *et al.* (1964) have studied the photobehavior of *Copilia* and their eye structure. It seemed important to us to learn more about the optics and imaging properties of such scanning eyes and also to see how their

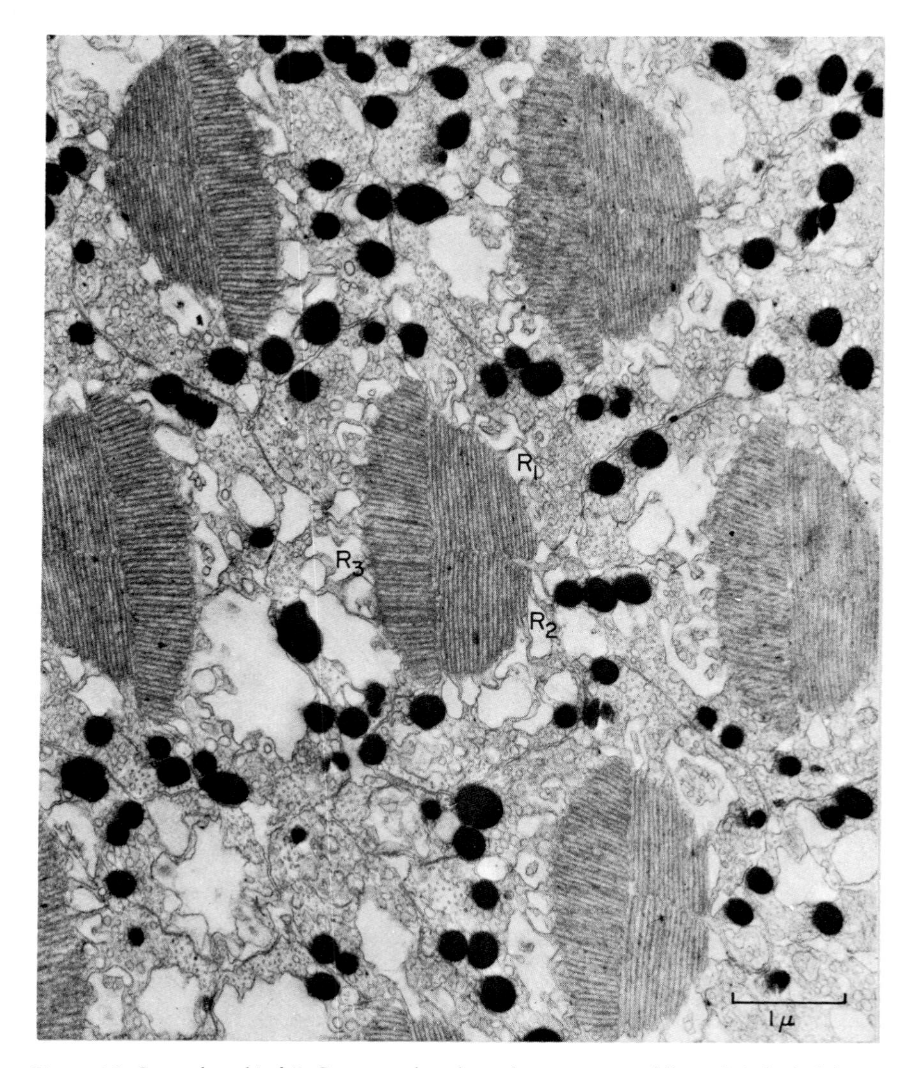

Figure 4.8. *Leptodora kindtii.* Cross section through many ommatidia and their rhabdoms.

photoreceptors are structured in comparison to other arthropod eyes (Wolken, 1958b).

The *Copilia* eye resembles an ommatidium of the compound eye with a corneal lens (biconvex anterior lens) and at some distance away the crystalline cone (posterior lens). Attached to the crystalline cone are the retinula cells which give rise to the rhabdomeres that form the rhabdom. The rhabdom lies in an L-shaped, orange-colored stem which is the only

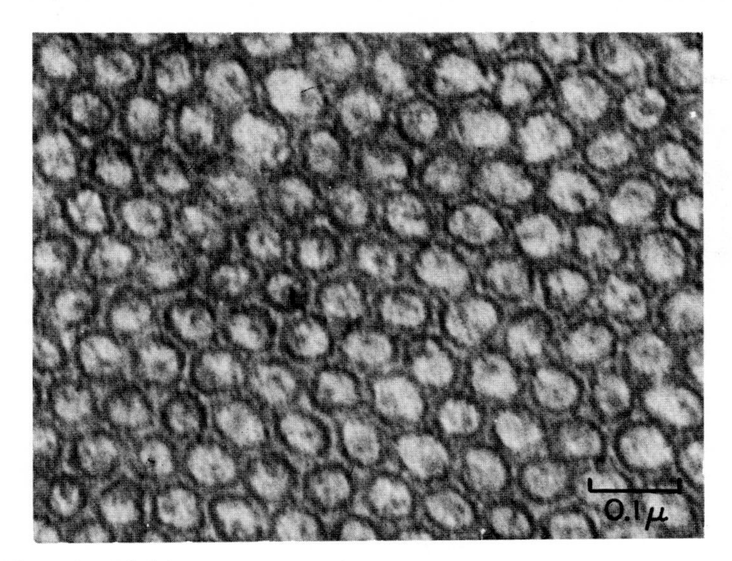

Figure 4.9. Rhabdomere, cross section of the microtubules. *Leptodora kindtii.*

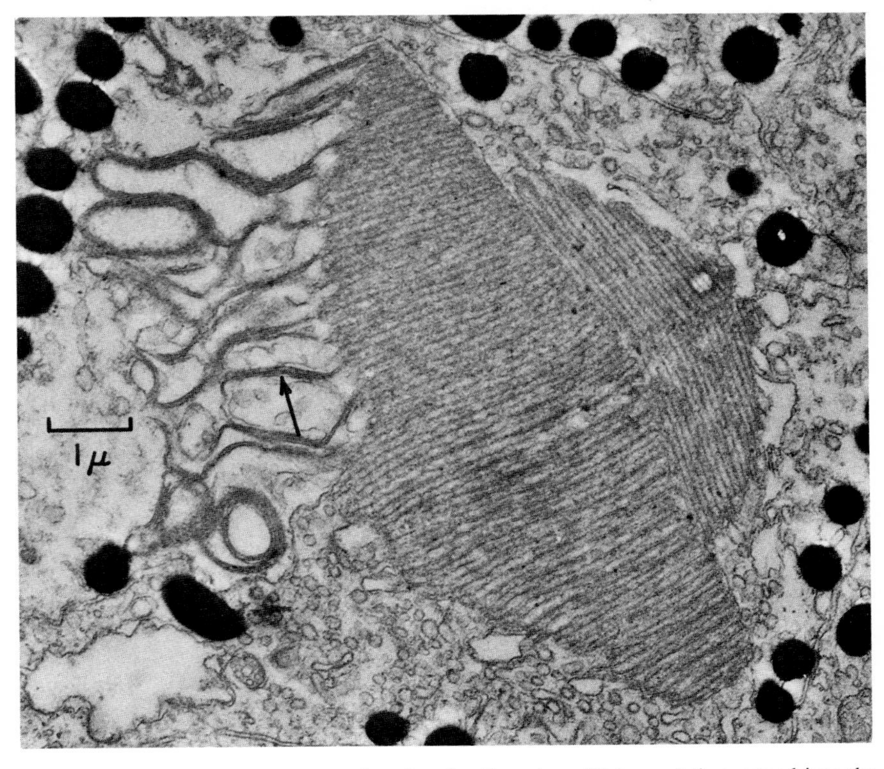

Figure 4.10. Rhabdom, cross section showing the microvilli (arrow) that extend into the cytoplasm of the retinula cell. *Leptodora kindtii.*

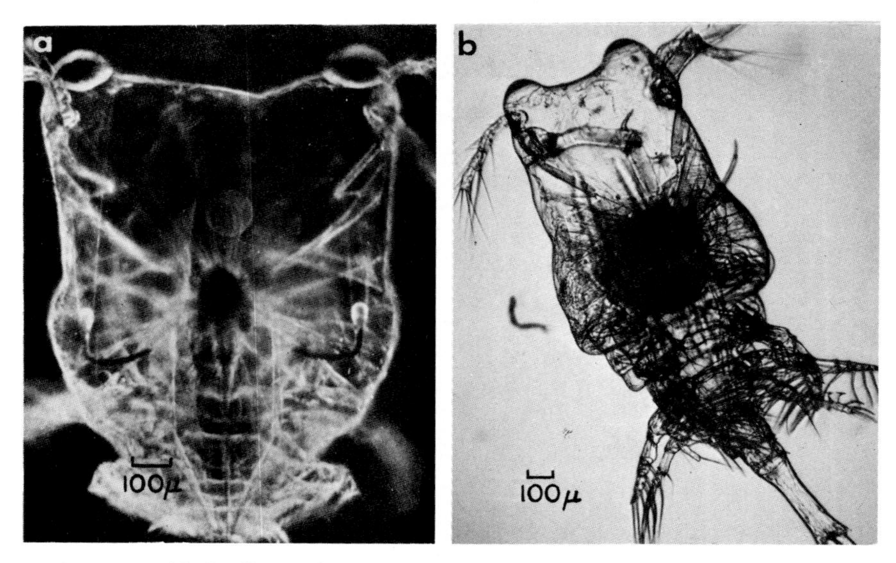

Figure 4.11. (a) *Copilia quadrata,* Mediterranean, dark field photomicrograph (courtesy of Dr. Neville Moray). (b) *Copilia mirabilis,* Caribbean.

pigmented part of the body (Figures 4.13 and 4.15). It is this stem that oscillates back and forth in a saw-toothed pattern, varying from about 1 scan/2 seconds to 5 scans/second (Gregory *et al.,* 1964). The stems from both eyes move synchronously and rapidly toward each other, then separate more slowly. Gregory (1966) has likened such scanning to a television camera: "It seems that the pattern of dark and light of the image is not given simultaneously by many receptors, as in other eyes, but in a time-series down the optic nerve, as in the single channel of a television camera."

In *Copilia quadrata,* the retinula cells lie directly behind the crystalline cone and are followed by the rhabdomeres that comprise the rhabdom. The rhabdom measures 11×17 μ and is completely surrounded by pigment granules. It extends about 60 μ in length from the retinula cells to the bend of the stem. Only five rhabdomeres (R_1–R_5) can be identified in the rhabdom (Figure 4.13d). One of them (R_1) is an asymmetric rhabdomere located at a nodule on the side of the stem facing the brain and lying at the base of the crystalline cone (Figure 4.13c). The asymmetric rhabdomere (about $1.7 \times 1.7 \times 7$ μ) appears to be at nearly 45° with respect to the stem. Rhabdomeres R_2–R_5 are ellipsoids measuring 1×2 μ and are about 58 μ in length. These rhabdomeres lie with their longest dimension parallel to the stem. The rhabdomeres are closely associated with mitochondria, and structures resembling nerve vessicles are also found in

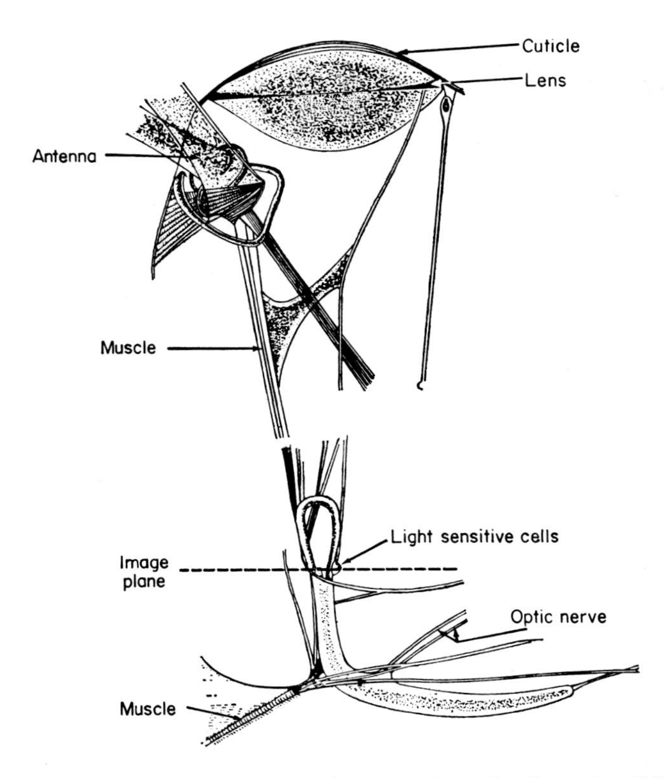

Figure 4.12. The structure of the *Copilia* eye as drawn by Grenacher (1879).

this region (Figures 4.13c,d and 4.14). Rhabdomeres R_1–R_3 are separated by screening pigment granules, whereas rhabdomeres R_4 and R_5 are not. The rhabdomere fine structure is that of packed microtubules about 500 Å in diameter (Figure 4.14), which is similar to that already described for the arthropod rhabdomeres.

The *Copilia* eye can be considered analogous to the superposition-type ommatidium in which the crystalline cone lies at some distance from the corneal lens. In addition, the crystalline cone forms a convex interface with a fluid of lower refractive index (see Figure 4.13). The structure of the crystalline cone resembles that of a cornea (Figure 4.13b). The concentration of this material varies across the diameter, the greatest concentration being in the center. Therefore, the crystalline cone would have the properties of a lens. The *Copilia* eye with its corneal lens, L_1, and its crystalline cone, L_2, may then be considered a two-lens optical system in which the posterior lens is positioned a short distance in front of the rhabdom (Figure 4.15b).

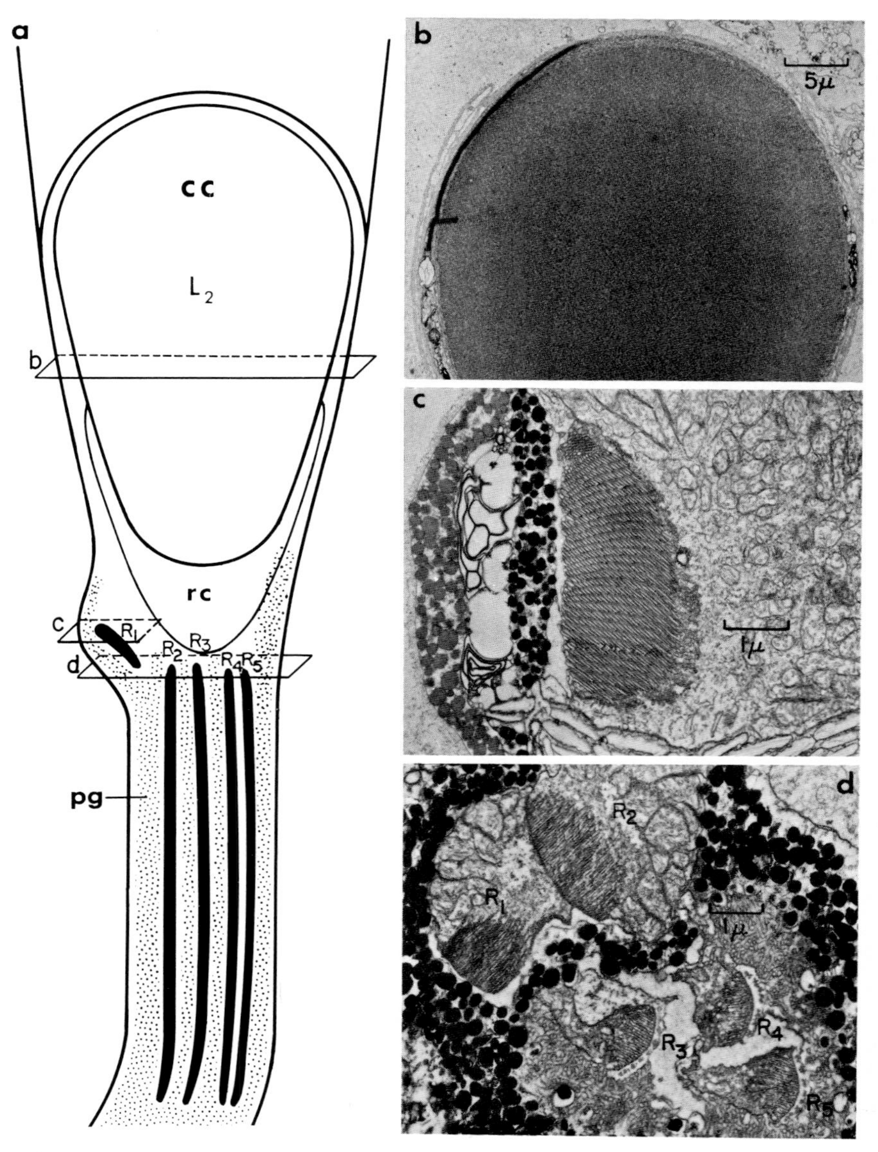

Figure 4.13. The *Copilia* eye. (a) Schematic, longitudinal view of the crystalline cone and rhabdom. cc(L_2), crystalline cone; rc, retinula cells; pg, pigment granules; rhabdomeres R_1–R_5 that form the rhabdom. Rectangles b, c and d show approximate areas of the electron micrograph sections. (b) Cross section of the crystalline cone (note the change in density towards the center). (c) Oblique section through the nodule showing the assymmetric rhabdomere, R_1. (d) Cross section of the rhabdom showing the five rhabdomeres (R_1–R_5).

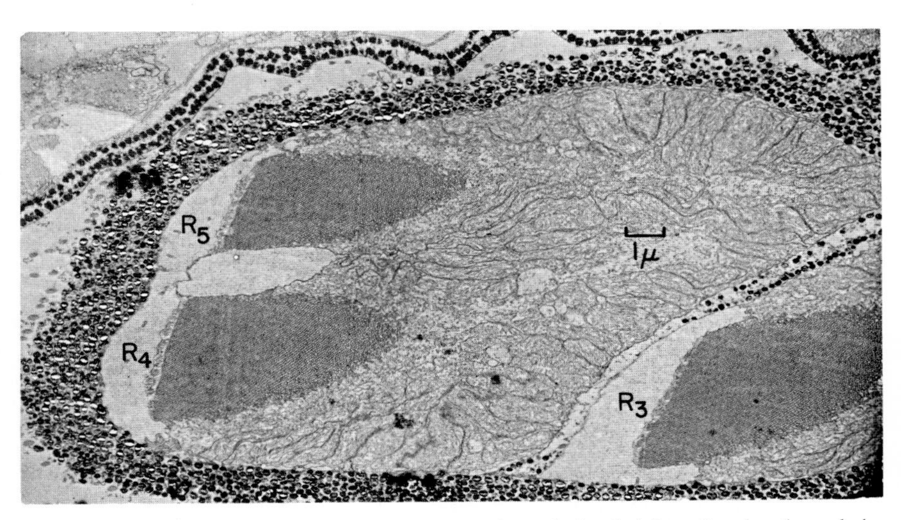

Figure 4.14. *Copilia quadrata*. Oblique section through the rhabdom showing three rhabdomeres (R_3–R_5) that have crystalline-like structures not found on rhabdomeres R_1 and R_2, (Figure 4.13d). Rhabdomeres R_4 and R_5 are interconnected and do not appear to be isolated by pigment granules at any level.

Eyes of both *Copilia quadrata* and *Copilia mirabilis* were measured to see if there were differences in the dimensions of their optical systems. These measurements indicated that the diameters and shapes of the corneal anterior lens (L_1) and the crystalline cone posterior lens (L_2) were approximately the same for both species. Also, the distances between lenses L_1 and L_2 were similar for both species.

In order to determine the imaging and optics of the *Copilia* eye, specimens were fixed, the lenses oriented as accurately as possible for measurements of their radii of curvature, and sections were then cut for light microscopy. The following measurements were obtained: the anterior lens, diameter 0.172 mm; radius of curvature of the front surface, 0.0994 mm; radius of curvature of the rear surface, 0.202 mm; thickness, 0.070 mm. The posterior lens diameter was 0.039 mm; radius of curvature of the front surface, 0.019 mm; radius of curvature of the rear surface, 0.0102 mm; and thickness, 0.0553 mm. The distance between the adjacent surfaces of the two lenses was 0.61 mm, and the distance from the rear surface of the anterior lens to the asymmetric rhabdomere was 0.67 mm.

The rhabdom is the open-type in which the rhabdomeres are separated (Figure 4.13d). A similar rhabdom structure is found in the insects *Musca domestica* and *Drosophila melanogaster* (Figure 3.8c). This

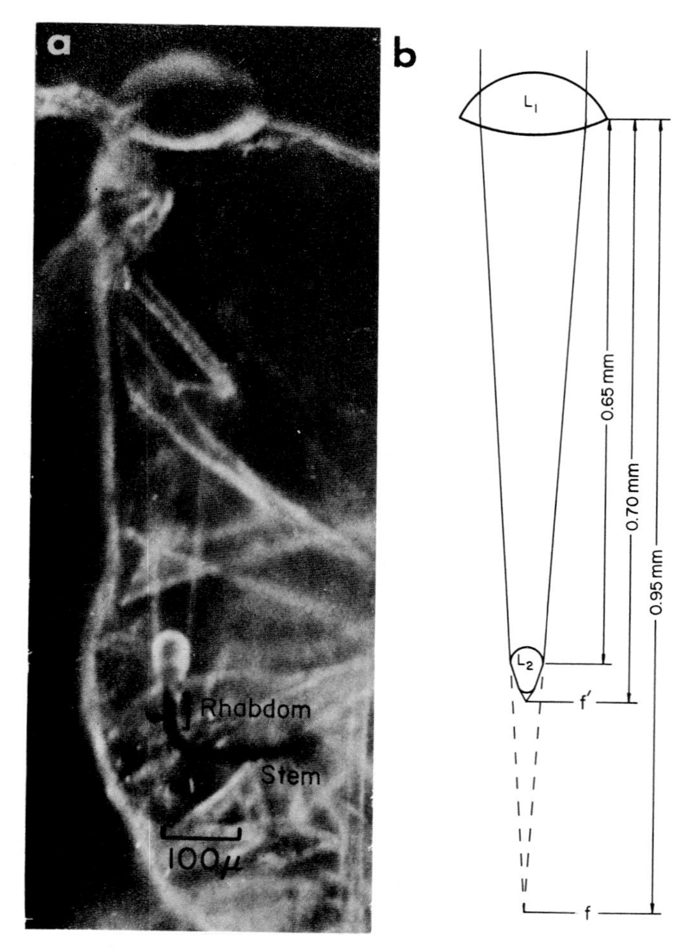

Figure 4.15. *Copilia quadrata*. (a) Dark field photomicrograph of the eye. (b) The optical system showing the positions of the corneal anterior lens (L_1) and the crystalline cone, posterior lens (L_2); f, focal point of the corneal (anterior) lens; f′, focal point of the total optical system.

structure differs from that of the fresh water crustaceans *Daphnia* and *Leptodora kindtii* that have a closed-type rhabdom in which the rhabdomeres are fused (Figures 4.2 and 4.7d). The open-type rhabdom is common to most diptera that navigate at high light levels, whereas most arthropods that navigate at low light levels have a closed-type rhabdom with the significantly greater effective cross section necessary for better light gathering efficiency. It was expected that *Copilia,* since it lives at a depth where the light level is near that of moonlight, would have a similar rhabdom for more light gathering. However, if the anterior lens (L_1) had a high relative speed, it could compensate for this less efficient open-type rhabdom.

To determine whether in fact the image is formed at the rhabdom, it is necessary to know the focal lengths of the anterior and posterior lenses. The focal lengths could be calculated by using our measurements for the radii of curvature of the lenses if we knew their respective refractive indices. Exner (1891) found that the focal length of the anterior lens in water was 0.93 mm. Grenacher (1879) had previously measured the distance from the corneal lens to the stem and found it to be 0.9–1.0 mm. These values are greater than those obtained for our specimens. For *Copilia quadrata* and *Copilia mirabilis* our measurements of the separation of the lenses was 0.65 mm (Figure 4.15), which is in agreement with Gregory *et al.* (1964). Therefore, we cannot assume that the anterior lenses in our specimens have the same focal length as that found by Exner (1891). Since we were not able to measure the focal length of the lenses in our specimens, we had to search for data that would permit its computation. We found that effective values of refractive index ranged from 1.42 as measured by Walls (1942) to 1.5 as measured by Kuiper (1966). If we take a value of 1.42 for the anterior lens, it would have a focal length of 0.98 mm; if we take a value of 1.50, the focal length would be 0.52 mm. Since the distance to the rhabdom is 0.69 mm, a refractive index of 1.46 for the corneal lens would place the image directly on the rhabdom. However, this does not take into account any refractive power of the posterior lens.

To examine how this optical system would work, we took a value of 1.425 for the refractive index of the anterior lens, a value which gave for this lens a focal length of 0.93 mm (i.e., the focal length measured by Exner). Using this focal length and the known position of the rhabdom, we found that the focal length of the posterior lens that would be necessary to place the image at the rhabdom level was 0.128 mm. The anterior lens alone has a relative speed or focal ratio of 5.5 : 1 but, when the posterior lens is taken into consideration, the focal ratio changes to 2.5 : 1, or an increase in light collecting efficiency of five times.

In order to see how efficient this optical system would be, we constructed a holder for a 15 mm diameter and 25 mm focal length Hastings triplet lens, and we mounted the lens 17.5 mm in front of the film plane of a 101.5 mm × 127 mm (Burke and James) commercial view camera. The image was then focused on the film plane as formed by a combination of that lens with the regular camera lens (a 152.5 mm focal length lens with a focal ratio 6.8 : 1), and photographs were taken. With the camera lens alone, an exposure of 1/25 second at *f*/18 was required to record an image. However, when the second lens was introduced, the exposure had to be reduced to 1/200 second to record an image with the same density on the negative. Calculations indicated that the lens speed was increased from *f*/18 to *f*/5.6, or an increase of more than eight times in image brightness. The photographs also showed that the size of the image formed by the combination of these lenses was reduced to about one-third, but this is much less than the gain in image brightness. A similar optical system has been described for a focal reduction camera used on the large telescope at Yerkes Observatory, Williams Bay, Wisconsin (Gascoigne, 1968).

Although the *Copilia* eye is considered primitive in that its field scanning mode is slow, it has adapted to low light levels by evolving a comparatively "advanced" optical system (Wolken and Florida, 1969).

Molluscs

Next to the arthropods, the molluscs are by far the most numerous of the invertebrates. Included among the molluscs are clams, oysters, nautilises, squids, and octopi. Almost every kind of imaging eye is found among them, from the simple pin-hole eye of the *Nautilis* to the refracting eye of the *Octopus* (Lane, 1960).

The Octopus

Of the mollusc cephalopods, it is the *Octopus* that attracts our immediate attention. The octopus is a large muscular animal; many grow to 10 feet in length and have eyes that are unusually large for the body size. Of interest to us here is that all its behavioral responses are primarily visual ones. Also, the brain behind the eye is one of the most highly developed of any of the invertebrates. It is enclosed in a cartilaginous "skull" and is divided into 14 main lobes that govern different sets of functions. The optic lobes are the largest of all; one set controls the jet apparatus; another set, the memory (Young, 1962).

The *Octopus* uses both monocular vision and binocular vision. In monocular vision the eyes face in opposite directions with their long axes roughly parallel. In binocular vision the eyes are switched slightly forward relative to the body. Therefore, the *Octopus* has a complete field of view of 360°. The *Octopus* can learn to distinguish one shape from another, even when the only difference between the two is their orientation. The *Octopus* can also achieve color matching with its surroundings, producing the pattern that it sees through expansion and contraction of the chromatophores in its skin (Packard and Sanders, 1969).

All of these visual responses are not completely understood. However, according to J. Z. Young (1960, 1962, 1964) the *Octopus* is an ideal animal for studies of visual acuity, eye–brain relationships, and learning processes.

I could not help but wonder what kind of eye this animal possessed. Considering that the *Octopus* (like *Copilia*) has a long history which shows it to be among the oldest species known, I was curious to see how their eyes and retinae are structured.

The *Octopus* eye (Figure 4.16) resembles in physical organization the vertebrate eye (Figure 3.3). However, its simple lens is formed out of two halves joined together, the retina is not inverted as in the vertebrate eyes, and the photoreceptors are directly exposed to the incident light (Hess, 1943; Ramsey, 1952). Histological studies indicated that the retinal photoreceptors are not rods and cones as in the vertebrate eye, but are groups of rods that form rhabdoms similar to those described for the arthropod compound eyes (Grenacher, 1886; Patten, 1887). Grenacher (1886) recognized that the rhabdoms were formed from four retinula cells.

When I studied the retina of *Octopus vulgaris* and the closely related species, the cuttlefish *Sepia officinalis* with the electron microscope, I found in agreement with Grenacher (1886) that each rhabdom consists of four radially arranged rhabdomeres (Wolken, 1958b).

In a cross section through the rhabdom, four sides of the rhabdomeres are isolated by pigment cells containing screening pigment granules.

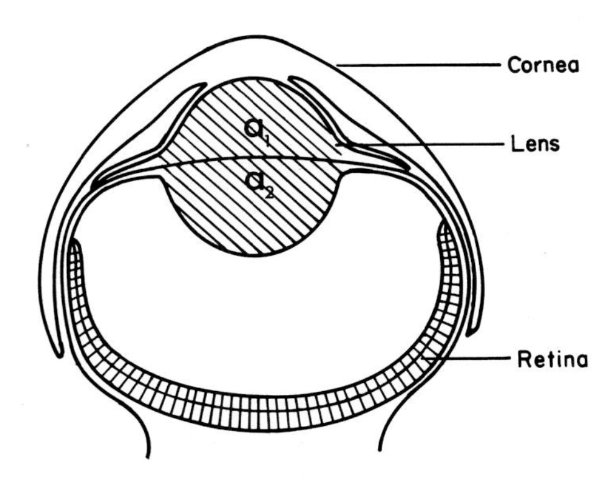

Figure 4.16. The mollusc cephalopod eye. (a_1 and a_2) Two halves of the lens.

These appear to migrate depending on the light intensity. In numerous cross sections and longitudinal serial sections the rhabdomeres measure about 70 μ in length and from 1 to 1.5 μ in diameter (Figure 4.17). In all the cross sections the lamellar structure is observed, whereas in all oblique and longitudinal sections the microtubular structure is seen (Figures 4.18 and 4.19). Here, as in the arthropod rhabdomeres, the microtubules are about 500 Å in diameter and are separated by a double membrane 100 Å thick. The *Octopus* retina is similar to a compound eye with ommatidia, as is schematized in Figure 4.20. The rhabdomeres form the rhabdom and their geometric structure is illustrated in Figure 4.18. Similar structures were found for the rhabdoms of *Sepia* and in Figure 4.19 for the squid (Zonana, 1961).

The retina of the *Octopus* is at least 1 cm^2 in area, which means that there would be about 2×10^6 rhabdomeres per eye. This number is roughly equivalent to the number of retinal rods of the vertebrate retina. However, the cephalopod retina differs fundamentally from that of vertebrates in that it contains no ganglion cells (Young, 1964).

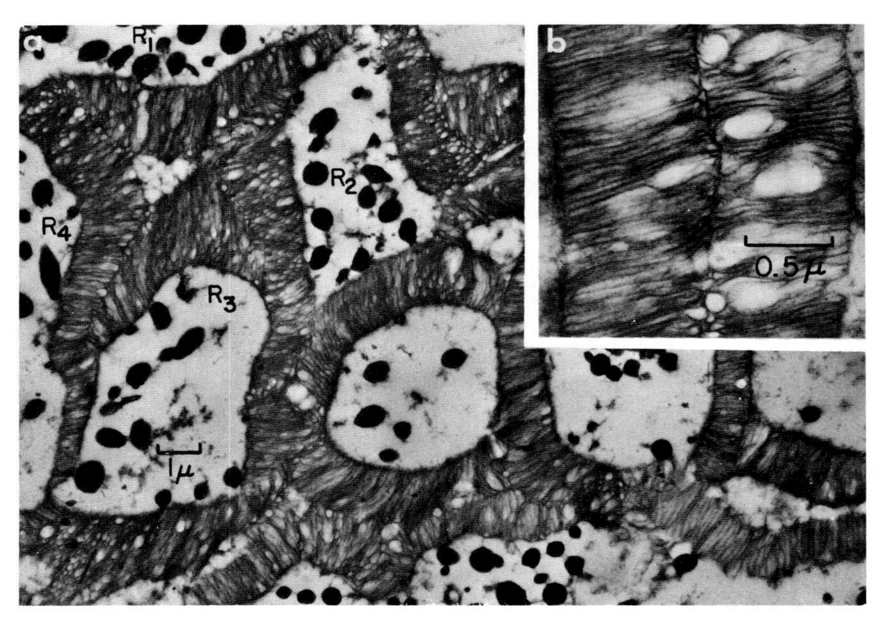

Figure 4.17. The cuttlefish, *Sepia officinalis*. (a) Oblique section to show the general geometry of the rhabdom (R_1–R_4) and structure of the rhabdomeres. (b) Two fused rhabdomeres, showing lamellar structure.

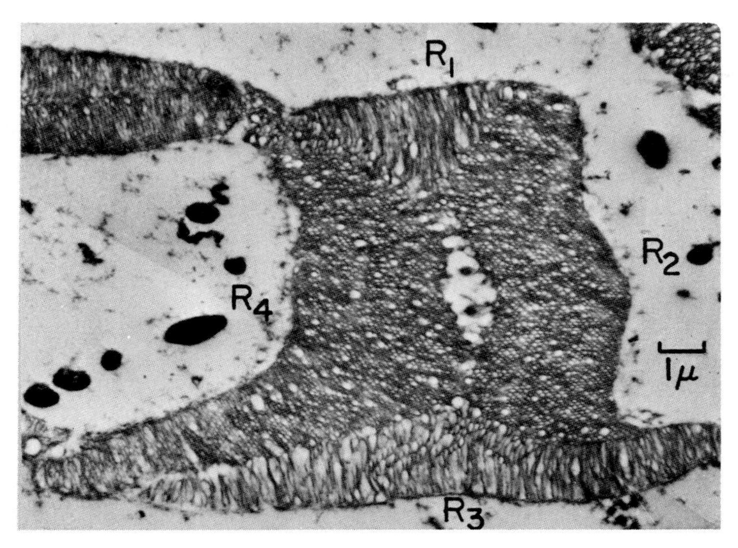

Figure 4.18. The cuttlefish, *Sepia officinalis*. Oblique section of rhabdom area showing two rhabdomeres with tubular structure (R_2 and R_4), and two with lamellar structure (R_1 and R_3).

The Rhabdom Structure

Despite the wide variety of invertebrate photoreceptor structures that have been revealed to us through electron microscopy, we are now better prepared to make some general observations about what we have seen.

As exemplified in the insects, crustacea, and molluscs just described, there are two structural arrangements for the rhabdomeres that form the rhabdom. One is the "open-type" rhabdom characteristic of the insect *Drosophila* (Figure 3.8) and the crustacean *Copilia* (Figure 4.13), in which the rhabdomeres project through a necklike part of their retinula cell into a comparatively large cavity. The other is a "closed-" or fused-type rhabdom as found in the cockroach (Figure 3.10), firefly (Figure 3.14), hornet (Figure 3.16), as well as the crustacean *Leptodora* (Figure 4.7), and molluscs *Sepia* and squid (Figures 4.18 and 4.19). In the fused rhabdom the retinula cells are modified to form wedge-shaped rhabdomeres that lie in close proximity to one another around a narrow axial cavity. However, exact information about how the fused rhabdom functions in visual excitation is sorely lacking.

The two structural arrangements are schematically illustrated in Figure 3.9 and in Figure 4.20. All rhabdomeres contain microtubules of about

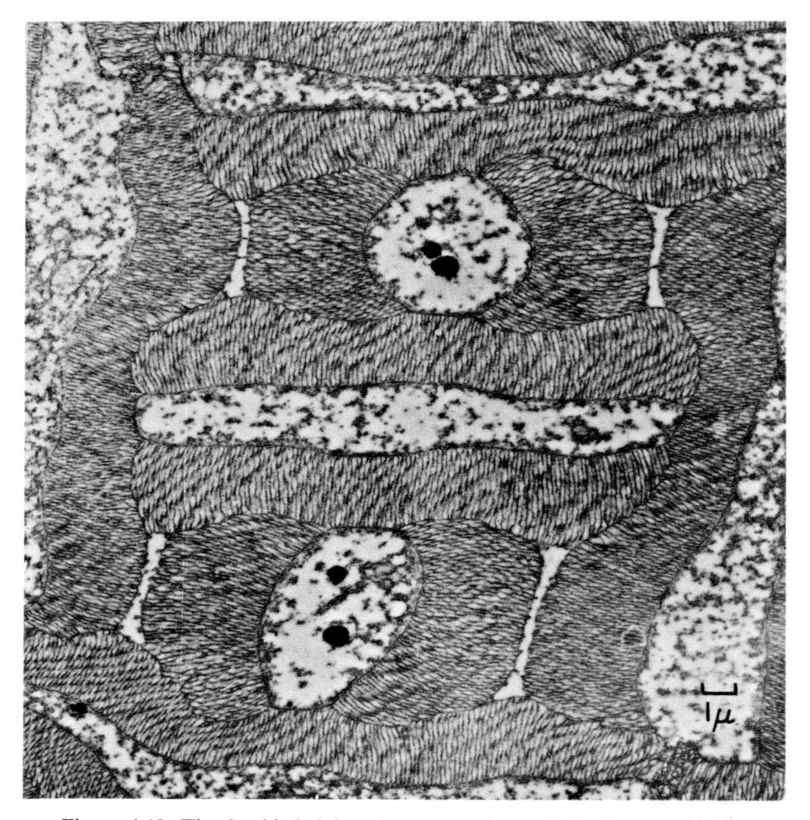

Figure 4.19. The Squid rhabdom (courtesy of Dr. H. V. Zonana, 1961).

500 Å in diameter with a double-membraned wall about 100 Å thick. In general, the microtubules of adjacent rhabdomeres are oriented perpendicular to each other, whereas those of opposite rhabdomeres are parallel.

Electrophysiological measurements indicate that there are also two physiological types of eyes, a "slow-type" eye characterized by a negative monophasic potential which is dependent upon the state of dark-adaptation, and a "fast-type" eye in which the electroretinogram (ERG) is diphasic, the magnitude and the form of the potential being independent of the state of dark-adaptation (Autrum, 1950, 1958). It is of interest, then, that except for the adult dragonfly, all the arthropods we have studied having a closed arrangement for their rhabdoms possess a slow-type electrical response. On the other hand, all the dipterous and hymenopterous insects that have an open-type arrangement possess the

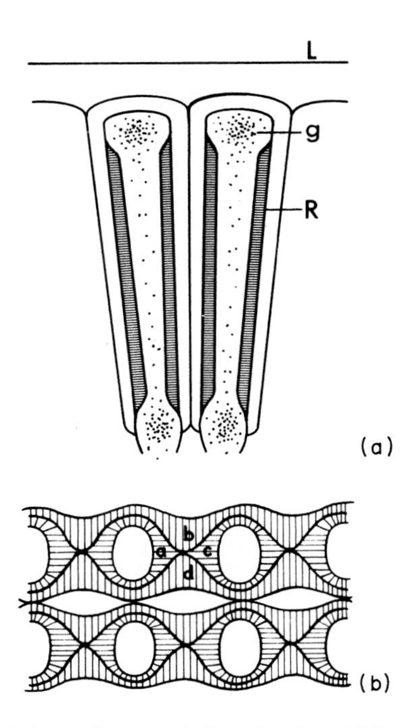

Figure 4.20. Schematic of the mollusc–cephalopod retina. (a) Longitudinal view through retina showing ommatidia and rhabdoms. (b) Cross section of the four rhabdomeres forming the rhabdom. L, Limiting membrane; g, pigment granules; R, rhabdomere.

fast-type electrical response. The slow-type eye is more characteristic of nocturnal insects and arthropods that navigate at low light levels. In the fast-type eye, the rhabdom occupies only a small part of the volume. In the slow-type eye, characteristic of nocturnal insects that have a light-gathering problem, the rhabdom takes up a much larger part of the volume; for example, the volume of the rhabdom of the cockroach is about five times the volume of the *Drosophila* rhabdom (Wolken, 1966).

Polarized Light

As we emphasized during our discussion of protozoa and insects, there exists a strong relation between the behavior of an organism and the structure of its photoreceptors. We now hope to fit our study of crustacea and molluscs into a similar framework.

Arthropods exhibit the unique ability to analyze the direction of plane-polarized light, using it as a compass for orientation and navigation.

Von Frisch (1949, 1950, 1953) and Autrum and Stumpf (1950) inferred primarily from behavioral studies of honeybees that the polarized light analyzers were built into the arrangement and structure of the rhabdomeres that form the rhabdom. Von Frisch (1950, 1953) constructed a model to show that the rhabdom consists of eight triangular polarizing elements, each transmitting a quantity of light proportional to the degree of polarization. In this model, opposite pairs of rhabdomeres were to have their polarizers in a parallel orientation (Figure 4.21). The arthropod and mollusc photoreceptor fine structure revealed by electron microscopy, and the strikingly geometric arrangement of perpendicular and parallel microtubules that form the rhabdom, both strongly support Von Frisch's model (Fernández-Morán, 1956, 1958; Wolken, 1966).

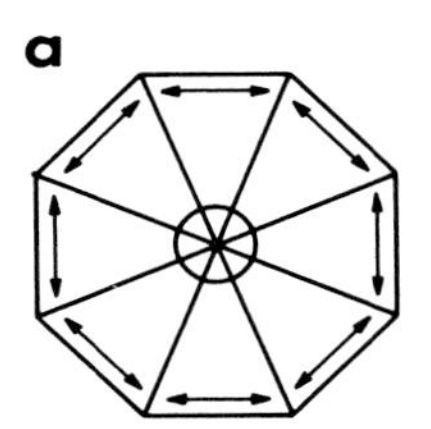

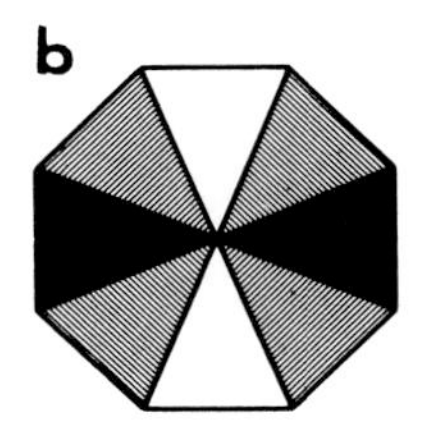

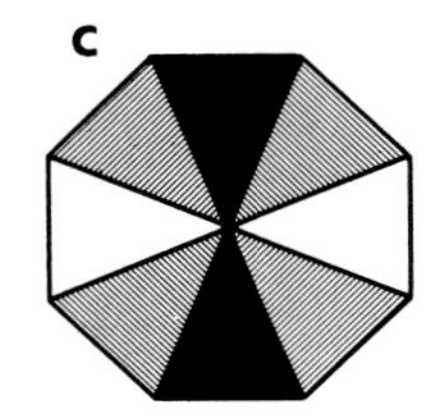

Figure 4.21. Rhabdomeres of rhabdom acting as an analyzer for polarized light. (a) The double arrows indicate the plane of vibration of the transmitted polarized light. (b) and (c) The rhabdom responds differentially to polarized light. Opposite rhabdomeres are stimulated to the same extent but adjacent ones are stimulated differently (after von Frisch, 1950, p. 107).

However, there are several crucial considerations which must be experimentally determined before we can conclude that the rhabdom fine structure is in fact the analyzer for polarized light. We must establish whether there is more than one pigment responsible for the organism's visual sensitivity. If there is more than one pigment, then we must discover the spectral and chemical characteristics of each as well as the functional relationships among them. In Chapters V and VI we will take up the nature of these visual pigments and their possible orientation within the retinal rod lamellae and rhabdomere microtubules.

The structural arrangement of the invertebrate rhabdom raises a number of interesting possibilities about the evolution of the vertebrate retinal rods. In the development of the retina the retinal rods form what may be considered a rhabdom-like array of receptors. This is seen more clearly in the amphibian retinal rods.

The possibility that there is an evolutionary structural and chemical basis for all visual photoreceptor systems, from the invertebrates to the invertebrate eye, will be considered in Chapters V, VI, and VII.

V. THE VERTEBRATE RETINAL PHOTORECEPTORS

One might wonder why, in the midst of our discussion of invertebrate photoreceptors, I have turned to the vertebrate retina. The reason is that before we attempt in Chapter VI to describe the experimental results of invertebrate visual pigments, we must find a basis for our analyses. This can be found in our extensive knowledge of vertebrate visual systems.

There is good reason to believe that throughout the animal world the kinds of pigment molecules and receptor structures involved in photosensitivity are similar. If this is true then we must expect certain features of vertebrate visual chemistry to repeat themselves in the invertebrates. With these considerations in mind, let us take a brief but very necessary look at the vertebrate retina, its visual pigments and photochemistry.

The Retinal Rod

The vertebrate retina consists of nine cell layers closely attached to the pigment epithelium (Figure 5.1). The nervous cell layers are the rod and cone cells, the bipolar cells, and the ganglion cells. Next to the retina is the *choroid* coat, a sheet of cells filled with black pigment which absorbs extra light and prevents internally reflected light from blurring the image. Toward the center of the human retina there is a depression, the *fovea,* which is the fixation point of the eye where vision is most acute. It contains mostly cones.

In the human retina (Figure 3.3) there are about 1×10^8 retinal rods and 7×10^6 retinal cones (Figure 5.2) of which 4×10^3 are in the foveal area. The rods become more numerous as the distance from the fovea increases. The fovea and the region just around it, the *macula lutea,* are colored yellow; they contain a plant carotenoid, a xanthophyll (Figure 1.6).

The rods and cones are differentiated specialized structures of the retinal cells. Each has an inner segment and what appears either as a rod- or cone-shaped outer segment which contains all of the photosensitive visual pigment (Figures 5.2, 5.3, and 5.5). Over a century ago, Schultze (1866) observed that when the retinal rods were exposed to light they disintegrated into plates. However, Schultze's observations had to await the developments of the polarizing microscope and the electron microscope before they could be confirmed.

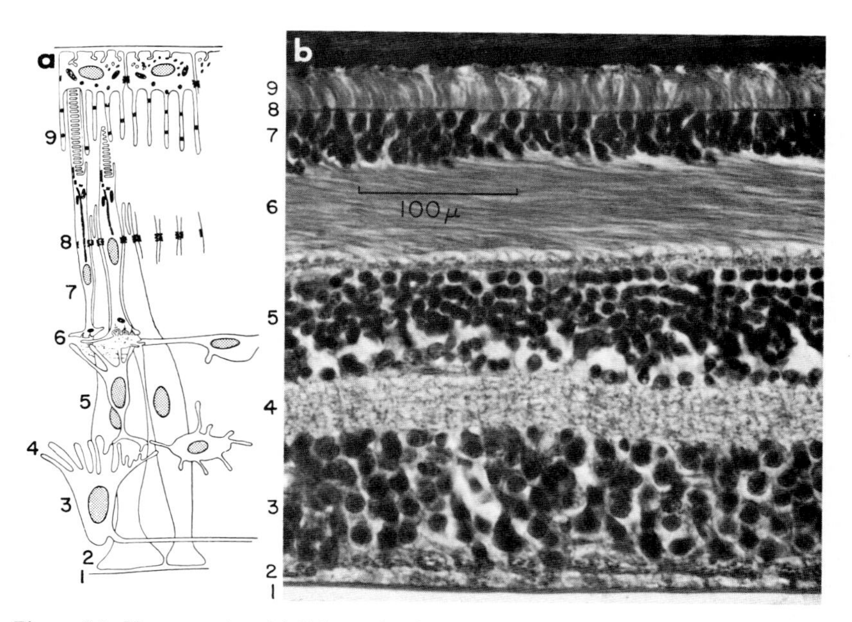

Figure 5.1. Human retina. (a) Schematic of the various cell layers (from Cohen, 1963b). (b) The retina. 1, Internal limiting membrane; 2, nerve fibers; 3, ganglion layer; 4, inner molecular layer; 5, inner nuclear layer; 6, outer molecular layer; 7, outer nuclear layer; 8, external limiting membrane; 9, rods and cones.

With the polarizing microscope Schmidt (1935, 1937, 1938) was able to show that the rod outer segments were birefringent—indicating that they were highly ordered structures. He observed that when the rods were extracted with organic solvents to remove the lipids there was a reversal in sign of the birefringence. To account for the observed optical changes Schmidt suggested that the lipid molecules would lie parallel to the axis of the outer segment and that there would be nonlipid material arranged at right angles to the long axis. When the rods were treated with dilute acids or alkali the positive birefringence disappeared, accompanied by a lengthening of the rod to as much as ten times its original dimension. Further observations on the dichroism of the outer segment suggested that the visual pigment was present in the nonlipid regions of the rod. With the rods free from lipid, Schmidt was able to show a complete curve for birefringence as a function of the refractive index of the medium. These observations led him to postulate that the rod outer segments were laminated structures of lipids and proteins.

Sjöstrand (1949, 1953a,b), using the electron microscope, then showed that the guinea pig retinal rod outer segments were lamellae, approxi-

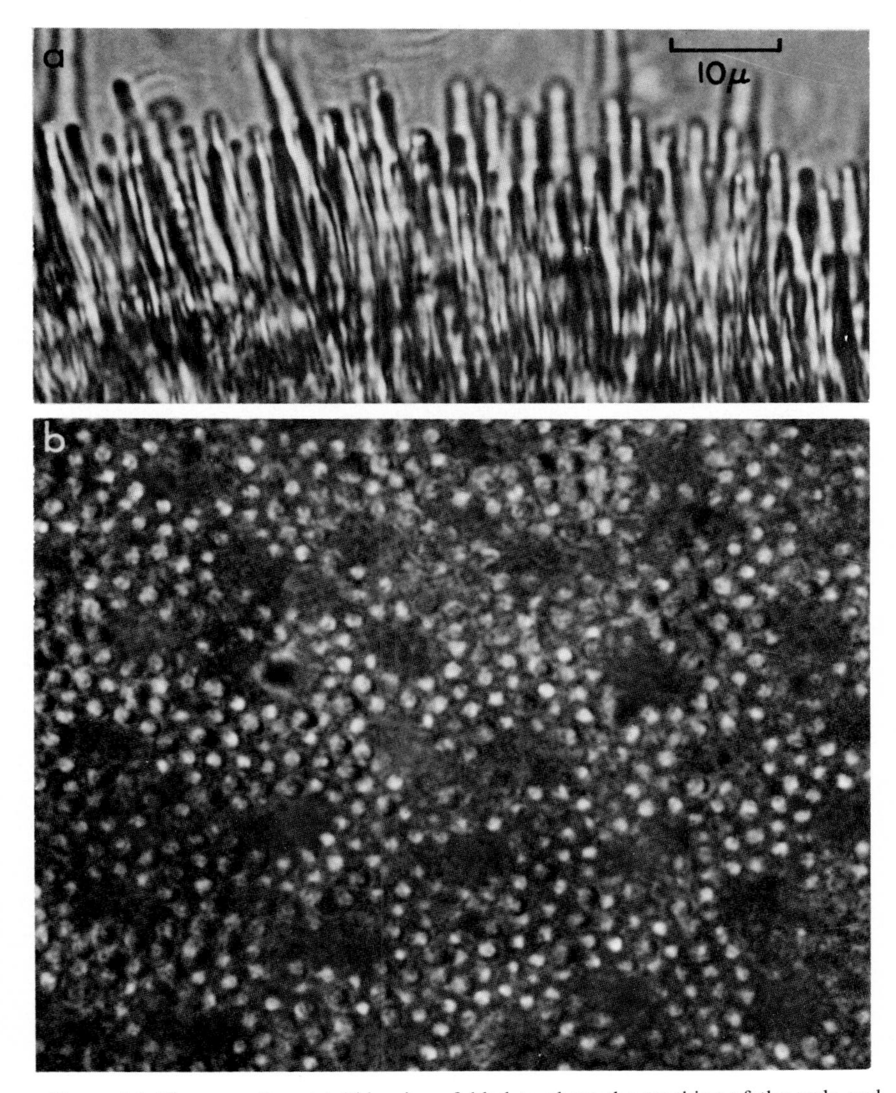

Figure 5.2. Human retina. (a) Side view, folded to show the stacking of the rods and cones. (b) Top view of the retina.

mately 2 μ in diameter, with edges 75 Å in thickness. Electron microscopic studies that followed clearly established that all vertebrate retinal rod outer segments (e.g., frog, perch, cattle, rabbit, rat, monkey, and human) are indeed double-membraned lamellae, of the order of 250 Å

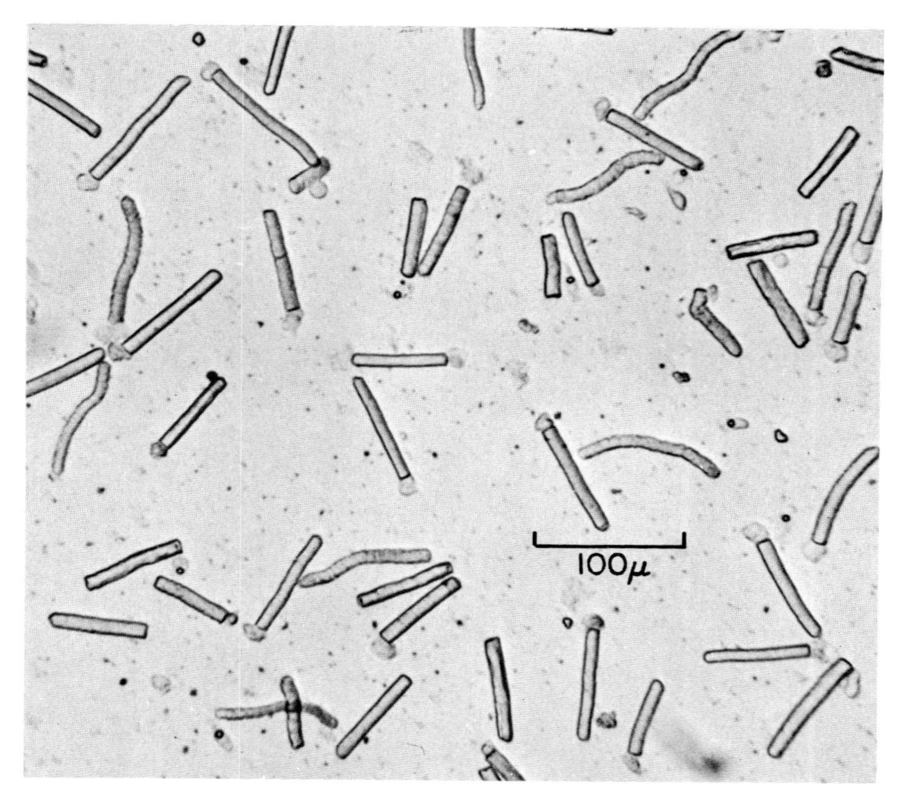

Figure 5.3. Frog retinal rod outer segments shaken from retina, suspended in Ringers solution.

in thickness, separated by less dense material of approximately equal thickness. Each membrane of the lamellae is from 50 to 75 Å in thickness (Cohen, 1961, 1963a,b, 1964; De Robertis, 1956; De Robertis and Lasansky, 1961; Dowling, 1965; Fernández-Morán, 1961; Missotten, 1964; Wolken, 1961a,b, 1963, 1966). Electron micrographs of the frog rod (Figure 5.4) and the bovine rod (Figure 5.5) illustrate this double-membraned lamellar structure.

The inner segments of the rods are not lamellar but are densely packed with mitochondria, as seen in Figure 5.6. The inner segment appears to be connected to the outer segment by a continuous membrane. Interconnection occurs by a thin filament that passes from the outer segment through the inner segment (Figure 5.5). Embryologically, the rods and cones are thought to have derived from flagella, and in this respect it is

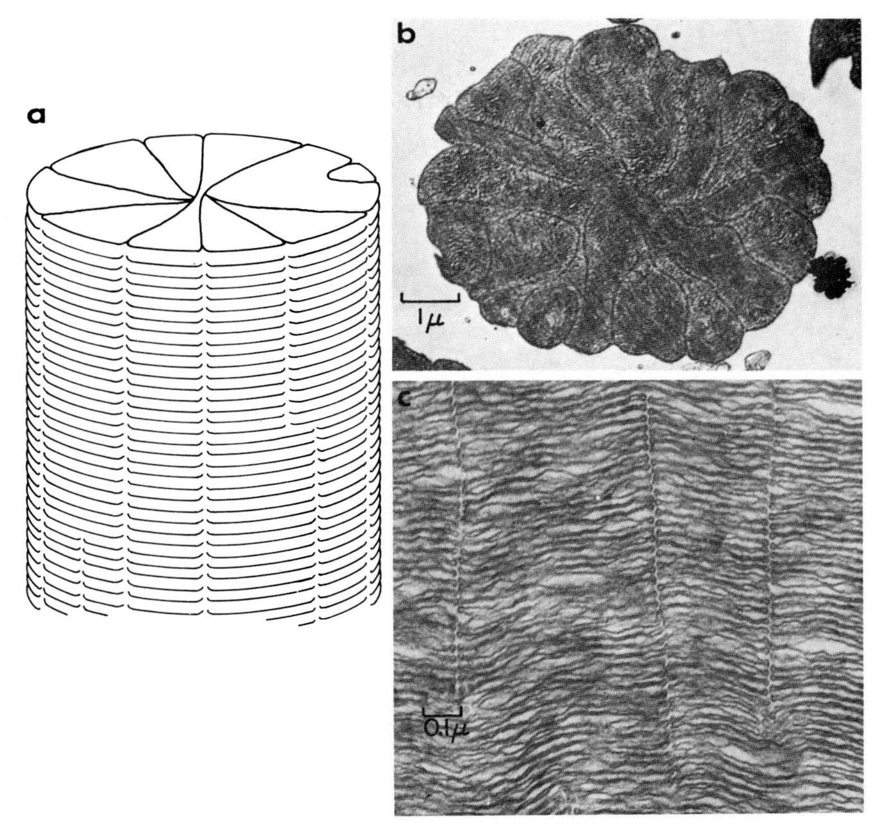

Figure 5.4. Frog retinal rod. (a) Schematic of the outer segment. (b) Cross section. (c) Longitudinal section, *Rana pipiens*.

interesting to note that communication between the segments is accomplished by a "flagellum" (De Robertis, 1956). The fine structure of this flagellum distinctly shows the characteristic nine fibrils found in the cilia and flagella of many plant and animal cells. For example, compare the cross section of a *Euglena* flagellum (Figure 2.6) with the connecting flagellum of the bovine retinal rod (Figure 5.5).

The close association of the flagellum with both highly specialized segments may be a crucial factor in understanding the functional chemistry of the rods. At one end of the rod we find a highly ordered matrix of photosensitive pigment well suited for light capture. At the opposite end we find a mass of tightly packed mitochondria whose enzymatic systems

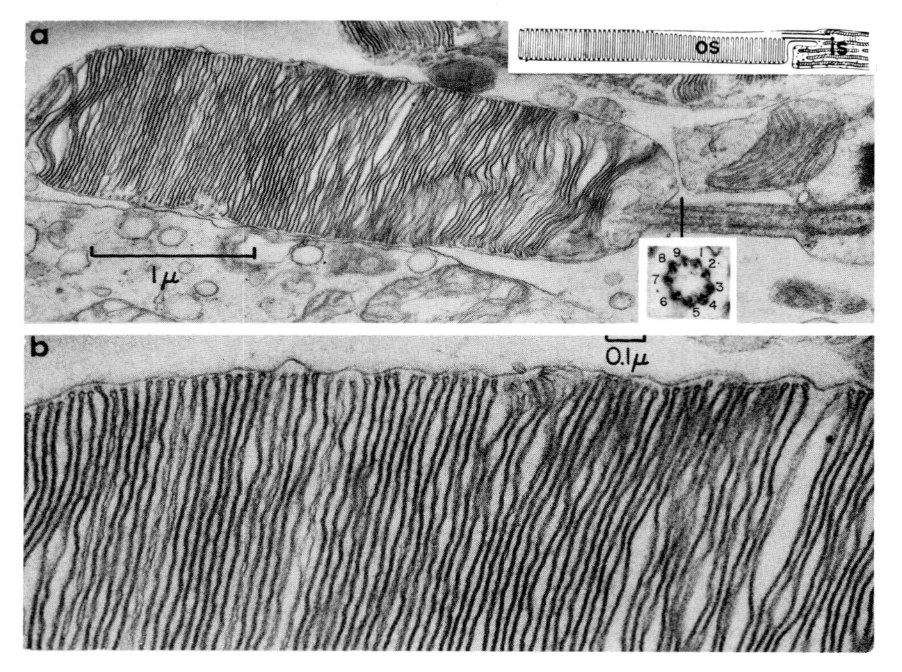

Figure 5.5. Cattle retinal rod (a) os, outer segment; is, inner segment. The electron micrograph (insert) of cross section of the cilium between outer segment and inner segment. (b) Enlarged section of the outer segment.

would provide for oxidation–reduction reactions and energy transfer (Figure 5.6).

The Amphibian Retinal Rod

Consider now the structure of the amphibian rod. The outer segment of the frog rod is an uncommonly large structure of about 60 μ in length and 6 μ in diameter (Figure 5.3). Thus it is easily observed with the light microscope and can be severed from the retina simply by shaking.

The isolated frog rods appear highly refractive and seem to be constructed of long, packed rodlets of about 1 μ in diameter. As soon as they are exposed to bright light the entire outer segment swells, and, as if subject to osmotic shock, begins to break transversely so that within a few minutes all structural identity is destroyed and its visual pigment is bleached (Wolken, 1961a).

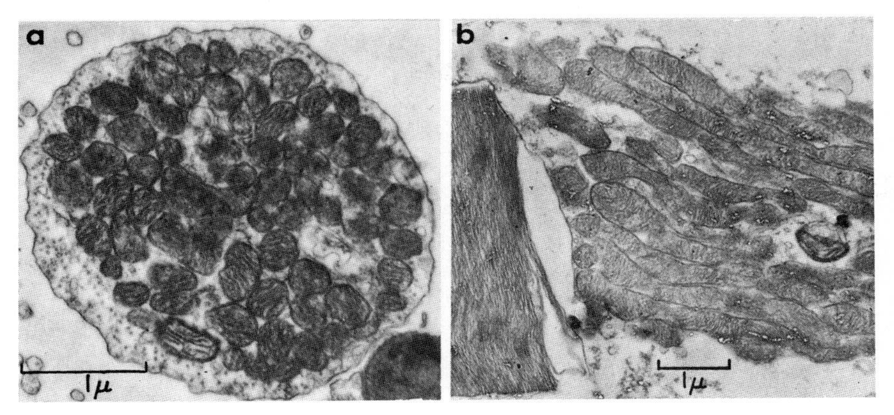

Figure 5.6. (a) Cross section of frog retinal rod inner segment. (b) Longitudinal section. Note the numerous mitochondria.

A cross-sectional view of freshly fixed and sectioned frog rods shows a cylinder with scalloped edges and fissures extending into the rod, so that it is divided into fifteen to twenty irregular, pie-shaped wedges (Figure 5.4). Longitudinal sections reveal that these wedges or rodlets produce further divisions within the structure of the rod (as shown schematically in Figure 5.4a).

In *Necturis,* the mud puppy, the rod outer segments are about twice the diameter of the frog rod and only one-half its length (12 μ in diameter and 30 μ long); they are the largest known among the amphibians (Wald *et al.,* 1963). Here too the outer segments are cut by a series of deep longitudinal grooves into an array of rodlets. There are from twenty to more than thirty such fissures making the rod in cross section appear scalloped, as in the frog rod (Figure 5.4b).

Thus the structure of the amphibian outer rod segment may be considered roughly analogous to that of a closed-type rhabdom, which we have already observed in certain arthropods (see, for example, Figures 3.10, 3.14, 3.16, and 4.8).

Perhaps what we are seeing here in the amphibian rod structure is a clue to an evolutionary link with the arthropods and molluscs, for all other vertebrate retinal rods are units of only 1 μ in diameter and show no such rhabdom-like structure. On the other hand, the resemblance between the closed-type rhabdom and the amphibian retinal rod may be the result of independent development and may have no evolutionary significance.

Vertebrate Visual Pigments

It will now be helpful to review briefly what is known of the biochemistry of the visual pigment system of vertebrate photoreceptors, the retinal rods and cones, before examining in detail the invertebrate visual pigments.

In the vertebrate eye, the retinal rods are for vision at relatively low light levels, and their maximal sensitivity is in the blue-green at about 500 nm. The cones are for bright light and color vision and their maximum spectral sensitivity lies more toward the red in the yellow-green at about 562 nm. (Figure 5.7). There occur small shifts in these absorption maxima depending on the animal species and their physical environment.

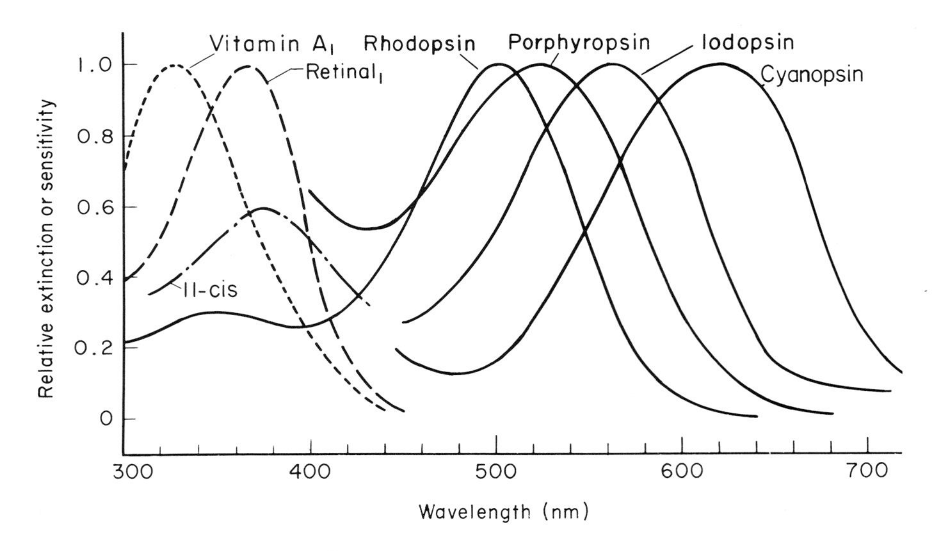

Figure 5.7. Visual pigments, absorption spectra (Wald, 1959; and Dartnall, 1957).

The photosensitive pigments upon which the visual threshold depends are either retinal$_1$ or retinal$_2$. In the revised nomenclature vitamin A is called *retinol,* vitamin A acid is *retonic* acid, and vitamin A aldehyde, or retinene, is *retinal*—so named because it was discovered in the retina. The extracted visual pigment complexes are identified primarily by their absorption spectra to be either *rhodopsin* (retinal$_1$ + rod opsin), or *porphyropsin* (retinal$_2$ + rod opsin) for the retinal rods, and *iodopsin* (retinal$_1$ + cone opsin) or *cyanopsin* (retinal$_2$ + cone opsin) for the retinal cones (Figure 5.8).

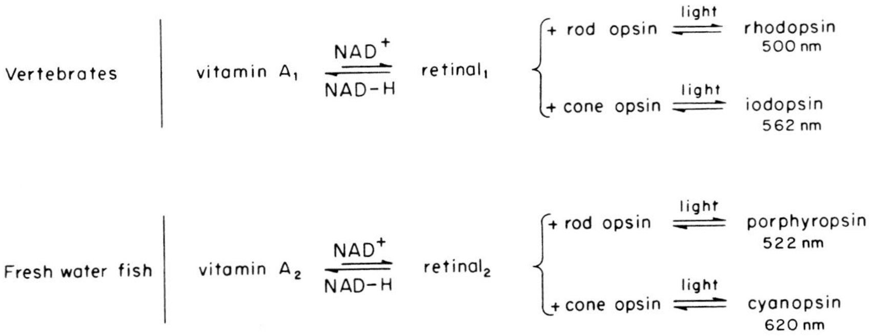

Vertebrates | vitamin A_1 $\underset{\text{NAD-H}}{\overset{\text{NAD}^+}{\rightleftharpoons}}$ retinal$_1$ { + rod opsin $\overset{\text{light}}{\rightleftharpoons}$ rhodopsin 500 nm; + cone opsin $\overset{\text{light}}{\rightleftharpoons}$ iodopsin 562 nm

Fresh water fish | vitamin A_2 $\underset{\text{NAD-H}}{\overset{\text{NAD}^+}{\rightleftharpoons}}$ retinal$_2$ { + rod opsin $\overset{\text{light}}{\rightleftharpoons}$ porphyropsin 522 nm; + cone opsin $\overset{\text{light}}{\rightleftharpoons}$ cyanopsin 620 nm

Figure 5.8. Visual pigments in the rods and cones, derived either from vitamin A_1, or vitamin A_2; indicating their absorption maxima (Wald, 1959).

The importance of retinal as an intermediate in vitamin A metabolism was established when it was demonstrated that retinal$_1$ is vitamin A_1 aldehyde, while retinal$_2$ is vitamin A_2 aldehyde. Morton (1944) and Morton and Goodwin (1944) were the first to suggest and confirm that retinal$_1$ is rapidly converted to vitamin A_1 when it is administered orally, subcutaneously, or intraperitoneally. The conversion of retinal to vita-

TABLE 5.1
GEOMETRIC ISOMERS OF RETINAL

Isomerization around bonds	Nomenclature	Alternate name
9–10	9-*cis*	iso-*a*
11–12	11-*cis*	neo-*b*
13–14	13-*cis*	neo-*a*
9–10, 13–14	9,13-di-*cis*	iso-*b*
11–12, 13–14	11,13-di-*cis*	neo-*c*

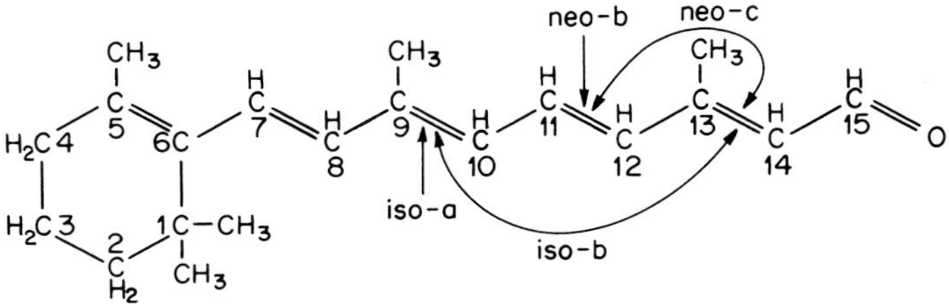

Figure 5.9. Retinal$_1$ indicating possible geometric isomers.

min A (retinol) is a reduction which occurs in the gut and in subcutaneous tissues. This reduction was a plausible explanation for the displacement of the absorption maximum (depending on the organic solvent) from around 328 nm, that of vitamin A_1, to about 370 nm, that of $retinal_1$ (Figure 5.7). The shift in the absorption spectrum can then be explained by an increase in the number of conjugated bonds from 5 to 6; if the terminal $—CH_2OH$ group of vitamin A (Figure 1.5) is replaced by a $—CHO$, the aldehyde group, this would provide the sixth conjugated bond (Figure 5.9).

$Retinal_1$ has been prepared by a number of workers. Morton and Goodwin (1944) were able to prepare the aldehyde by shaking vitamin A_1 concentrates dissolved in light petroleum ether with dilute aqueous potassium permanganate containing sulfuric acid. On chromatographic separation, a fraction was isolated which had an absorption maximum at 365–370 nm in saturated hydrocarbon solvents, and at 385 nm in chloroform. When reacted with antimony trichloride, the absorption maximum was at 665 nm. Hawkins and Hunter (1944) likewise synthesized $retinal_1$ from vitamin A_1, and Hunter and Williams (1945) obtained $retinal_1$ by the oxidation of β-carotene. This latter work is of special interest as it provides chemical proof for the conversion of β-carotene to vitamin A. $Retinal_1$ may also be an intermediate product in the transformation of β-carotene to vitamin A_1 *in vivo*.

$Retinal_1$ was synthesized in the course of synthesizing vitamin A_1 (van Dorp and Arens, 1947). Similarly, a procedure for the preparation of $retinal_1$ from vitamin A_1 was developed by Ball *et al.* (1948). In this method, vitamin A_1 is oxidized in light petroleum ether with manganese dioxide for 6 to 10 days in the dark; this yields about 80% $retinal_1$.

To further purify the $retinal_1$, it is fractionated chromatographically on an alumina column and then crystallized from petroleum ether at −72°C. After recrystallization, reddish-brown crystals are obtained which melt at 56.5° to 58°C. A final recrystallization from petroleum ether leaves large clusters of predominantly needlelike orange-red crystals that melt at 61°–62°C. The crystalline $retinal_1$ dissolved in cyclohexane shows a single maximum absorption peak at 373 nm, and when reacted with the Carr-Price Reagent, antimony trichloride, an absorption maximum at 664 nm is obtained.

Wald (1938, 1939) was the first to demonstrate the presence of $retinal_2$ in the frog retina, and $retinal_2$ was later crystallized by Salah and Morton (1948). The absorption maxima of crystalline $retinal_2$ was found to be 385 nm in cyclohexane, 388 nm in light petroleum ether, 395 nm in ethanol, and 406 nm in chloroform. $Retinal_2$ also reacts with antimony trichloride, showing an absorption maximum near 703 nm.

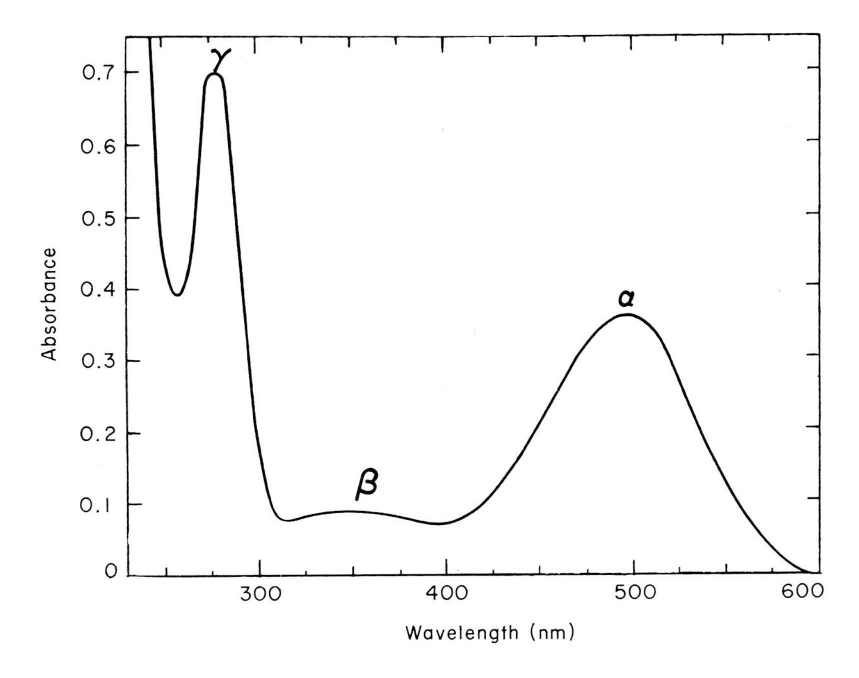

Figure 5.10. Cattle rhodopsin absorption spectrum (Bowness, 1959, p. 309).

The contents of the retinal rods are not easily solubilized and they require the use of special solubilizing agents to extract the visual pigment. Extraction of visual pigments is an arduous task, for it requires a large number of eyes, dissection, solubilization, and fractionation in the dark and in the cold. The procedures for extraction and the use of various solubilizing agents are described by Dartnall (1957). A most effective extractant and one that is most commonly used is digitonin (a digitalis glycoside). After extraction, relatively pure rhodopsin can be obtained (Figure 5.10) by the use of column fractionation (Bowness, 1959; Heller, 1968, 1969).

The Photochemistry of Rhodopsin

The mechanism of the visual process in the vertebrate retinal rod has been studied thoroughly by Wald (1953–56, 1959, 1961), Dartnall (1957, 1962) and their associates. The isolated visual pigment, rhodopsin (Figure 5.10), bleaches upon exposure to light, yielding retinal and its carrier protein *opsin*. The bleaching of frog rhodopsin by light can be followed by spectroscopy (Figure 5.11). Curve 1 is unbleached rhodopsin and curves 2–7 show the displacement of the major rhodopsin peak from close to 500 nm to that of all-*trans*-retinal near 370 nm. Retinal recombines with opsin in the dark to form rhodopsin. The reaction is spontaneous, and therefore opsin may be looked upon as a retinal-trapping enzyme removing free retinal from the mixture and causing the production of additional retinal from vitamin A to maintain the necessary equilibrium.

Vitamin A and retinal are polyene chains (Figures 1.5, 1.6, and 5.9) that exist in a number of different geometrical configurations corresponding to the possible *cis–trans* isomerizations around the double bonds of these molecules (Figure 5.9, and Table 5.1). There are five possible geometric isomers of retinal corresponding to rotation about the 9-10

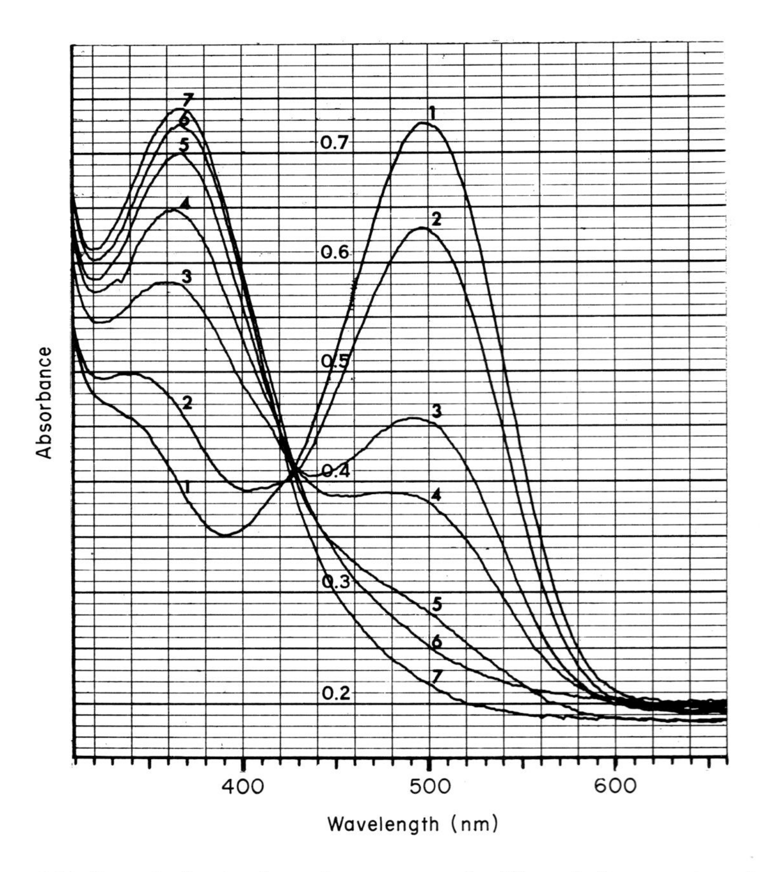

Figure 5.11. Frog rhodopsin absorption spectrum (in 4% tergitol), curve 1, and light bleaching spectra, curves 2–7.

carbon bond, the 11-12 bond, the 13-14 bond, the 9-10 and 13-14 bonds, and the 11-12 and 13-14 bonds. The retinal produced by bleaching rhodopsin has always been found to be in the all-*trans* form. However, to resynthesize rhodopsin from all-*trans*-retinal, it was found that retinal must be isomerized to the 11-*cis* (neo-b) form in order to recombine with opsin.

The existence of the 11-*cis* isomer was considered improbable, because the steric interference between the methyl group at carbon 13 and the hydrogen at position 10 would prevent the molecule from becoming entirely planar. Professor Wald and his associates were surprised to find that the functional isomer of retinal was in fact the 11-*cis* configuration of retinal. Because of its hindered configuration, the 11-*cis* form is the least stable of the possible isomers; it is the most easily formed upon irradiation, and the most sensitive to temperature and light. Wald feels that this unstable form of retinal, rather than a more stable form, helps explain its presence in the rhodopsin molecule. What could be more appropriate than a molecule that is very unstable in the light, but extremely stable in the dark.

The bleaching of rhodopsin to release retinal from opsin does not proceed directly, but in a series of intermediate steps (Figure 5.12). These steps have been recently reviewed by Abrahamson and Ostroy (1967) and the intermediates of bleaching by Guzzo and Pool (1969). The initial step is the isomerization of 11-*cis*-retinal to all-*trans*-retinal, resulting in a release of energy due to its transformation to a more stable form. This single step results in the conversion of rhodopsin to lumirhodopsin. It is followed immediately by a thermal rearrangement of the opsin molecule that produces metarhodopsin (Figure 5.13).

Light-induced changes in rhodopsin are not normally reversible, but at temperatures between $-196°$ and $-140°C$ the pigment is reversibly photochromic (Yoshizawa and Wald, 1963); rhodopsin (λ max 500 nm) $\rightleftharpoons$ prelumirhodopsin (λ max 543 nm). Thus light of 440 nm favors the forward

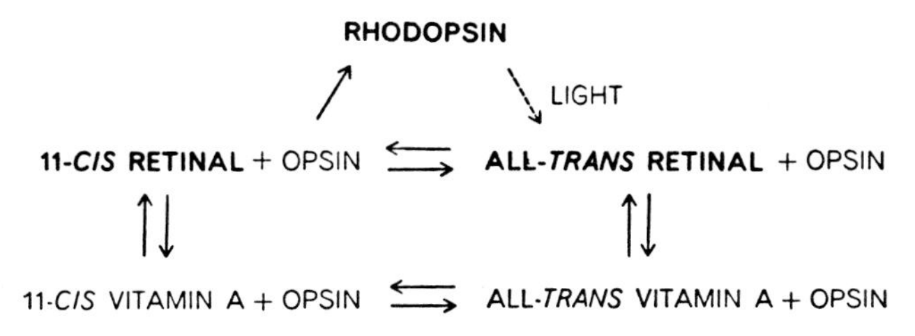

Figure 5.12. Photochemical events in vision.

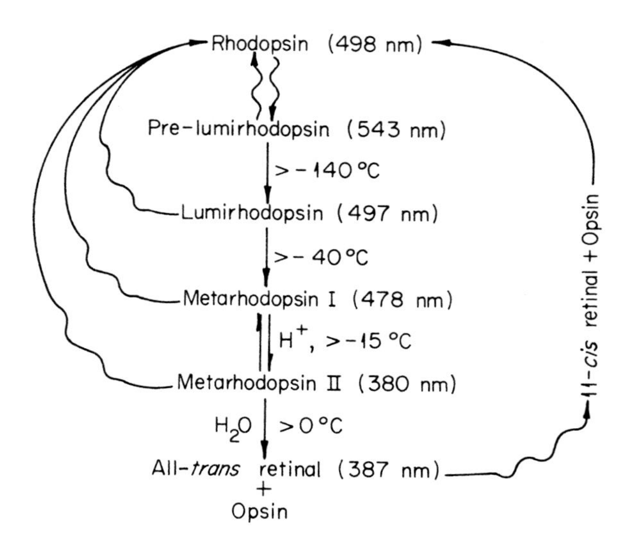

Figure 5.13. Rhodopsin bleaching. Photochemical events denoted by wavy lines, thermal (dark) reactions by straight lines (from Matthews *et al.,* 1963, p. 219).

reaction, while light of 600 nm drives the reaction to the left. This change apparently involves the isomerization of 11-*cis*-retinal to the all-*trans* form, but with little change occurring in the protein opsin as seen from its spectral maximum at 280 nm. As the temperature is raised, pre-lumirhodopsin changes to lumirhodopsin, presumably as the protein opsin "opens." By this time, excitation has occurred. Metarhodopsin then slowly hydrolysizes, except in some invertebrate retinas, to a mixture of all-*trans*-retinal and opsin.

The necessary light reaction is thus the isomerization of the 11-*cis*-retinal to the all-*trans*-retinal, resulting in the release of energy; all other reactions can be considered dark reactions.

The Retinal Cone Pigments and Color Vision

Let us turn now to the cone pigments and their reaction to color vision. It was noted earlier that chemical isolation has yielded only one photosensitive cone pigment, either *iodopsin* (λ max 562 nm) if retinal_1 is present, or *cyanopsin* (λ max 620 nm) if retinal_2 is present (see Figures 5.7 and 5.8).

How do these two pigments fit into our account of color vision in vertebrates? It may be recalled that there are two general theories of color vision, the *tricolor theory,* which arose from the work of Young (1802,

1807), von Helmholtz (1852, 1867), and Maxwell (1861, 1890), and the theory of Hering (1885, 1965). The tricolor theory asserts that there are three different pigments in the retinal cones with maximum absorption in the blue, green, and red regions of the spectrum. According to the theory, the brain computes yellow and white from green and red at high light intensities and white from blue at low intensities. In contrast, Hering's theory postulates that there are the following six basic responses, which occur in pairs: blue–yellow, red–green, and black–white. Excitation leading to any single response suppresses the action of the other member of the pair.

Recently, it has been possible to isolate the cones *in situ* and obtain their absorption spectra by microspectrophotometry. For example, frog cones were found to have general absorption throughout the whole of the visible spectrum with maxima near 430, 480, 540, 610, and 680 nm (Wolken, 1966). For the carp, cone peaks in the regions of 420–430, 490–500, 520–540, 560–580, 620–640, and 670–680 nm were found (Hanaoka and Fujimoto, 1957). In the goldfish, which belongs to the carp family, Marks (1963) found cones with absorption peaks at 455, 530, and 624 nm, and Liebman and Entine (1964) found absorption peaks at 460, 540, and 640 nm. For human and monkey foveal cones absorption peaks were found at 445, 535, and 570 nm (Marks *et al.,* 1964; and MacNichol, 1964). Wald (1964), Brown and Wald (1964), and Wald and Brown (1965) found for human cones absorption peaks near 450, 525, and 555 nm, which compare well with their psychophysical data for the spectral sensitivity of the human eye with peaks at 430, 540, and 575 nm. These studies indicate that for color vision there are at least three different spectrally absorbing cone pigments consistent with the psychophysical data; one for sensing *blue,* one for *green,* and one for *red.*

Molecular Structure of the Retinal Rod

The fundamental problem in vision is to attempt to understand how the action of light on a visual pigment leads to a nervous excitation. Though we reviewed the photochemistry of visual pigments in solution, the excitation problem can have only limited meaning in that context alone. Thus it becomes necessary to go from the rhodopsin molecule in solution to the retinal rod itself, of which rhodopsin is the fundamental part.

Microspectrophotometry has proven extemely useful in obtaining *in situ* spectral information from a single retinal rod. In Figure 5.14, the spectrum from a single frog rod is shown with its bleaching spectra,

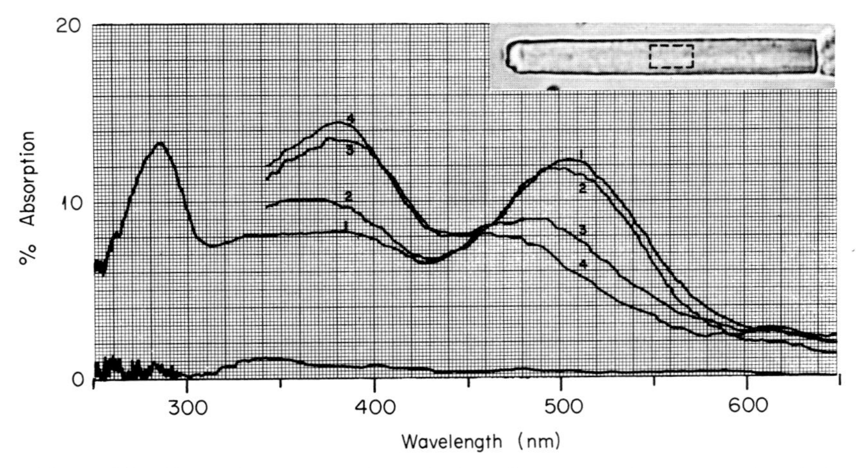

Figure 5.14. Absorption spectra obtained with microspectrophotometer, of an isolated frog rod, curve 1; curves 2–4, after bleaching with green light (500 nm) for 15 second periods. Note the shift in spectra from rhodopsin, curve 1, to that of retinal, curve 4. The ultraviolet peak for opsin near 280 nm does not shift. The insert shows the $4 \times 8\ \mu$ sample area.

curves 2–4 (compare Figure 5.14 with Figures 5.10 and 5.11). From these spectra (Figure 5.14, curve 1) an estimate for the number of rhodopsin molecules per rod can be calculated once we know the volume of the rod, its optical density, and the molar extinction coefficient of rhodopsin at 500 nm. Using our data we found that there were approximately 3×10^9 rhodopsin molecules per frog rod. From absorption spectra of rhodopsin extracted from a known number of rods, we found that the number of rhodopsin molecules per rod was 3.8×10^9 (Table 5.2). So we see that our estimates from both microspectrophotometry and solution studies are in excellent agreement.

TABLE 5.2
RETINAL ROD STRUCTURAL DATA[a]

Animal	Diameter, D (μ)	Thickness of dense lamellae, T (Å)	Number of dense layers per rod, n	Number of rhodopsin molecules per rod, N	Calculated cross-sectional area of rhodopsin, Å²	Calculated diameter of rhodopsin molecule, d (Å)	Calculated molecular weight, M[b]
Frog	5.0	150	1000	3.8×10^9[c]	2620	51	60,000
Cattle	1.0	200	800	4.2×10^6	2500	50	40,000[d]

[a]Taken from Wolken (1961a)
[b]Heller (1969) finds that the molecular weight of cattle and frog rhodopsin is 28,000.
[c]Microspectrophotometry of frog rod gives 3.0×10^9 rhodopsin molecules (Figure 5.14).
[d]Calculation based on a lipoprotein, density 1.1, gives a molecular weight of 32,000.

It is apparent from Figures 5.4 and 5.5 that the rod outer segments are comprised of double-membraned lamellae with a repeated unit of about 250 Å. Although the precise location of the rhodopsin, protein, and lipid molecules within these structures is not known, we can begin to build a structural model of the retinal rod. In order to do this, the geometry of individual rods—their length, diameter, number of lamellae and their dimensions—was determined from numerous electron micrographs (Wolken, 1956b). With these data and the previously determined rhodopsin concentration, we calculated the surface area each rhodopsin molecule could occupy (Wolken, 1961a).

The basis for this calculation was the assumption that the lamellae are double membranes of lipids and lipoprotein. The lamellae double layer is then structurally conceived as lipoprotein with the low molecular weight lipids occupying the interstitial spaces. It was further assumed that there is a monomolecular layer of rhodopsin molecules associated with the lipoprotein at the interfaces, and that these double membranes were separated by aqueous proteins. Both of these assumptions are supported by chemical analyses which show that, in general, the visual pigment accounts for 4–10%, protein for 40–50%, and lipids for 20–40% of the total outer rod segment weight.

The cross-sectional area A associated with each rhodopsin molecule is expressed by

$$A = \pi D^2/4P,$$

where D is the diameter of the retinal rod and P is the number of rhodopsin molecules in a single monolayer. In our calculation, P is replaced by $N/2n$, where N is the rhodopsin concentration in molecules per retinal rod and n is the number of double-membraned lamellae per rod. So the expression for the maximum cross-sectional area for each rhodopsin molecule is given by:

$$A = \pi D^2 n/2N.$$

Using this equation, the cross-sectional areas calculated for cattle and frog rhodopsin were 2500 and 2620 Å^2, respectively (Table 5.2), which means the diameter of the rhodopsin molecule should be of the order of 50 Å. This value agrees well with theoretical calculations, since a rhodopsin molecule, if symmetrical, would have a diameter of the order of 40 Å (Wald, 1954). Thus our figure for the available area implies that there would be sufficient space for all of the rhodopsin molecules to cover all of the lamellar surfaces of a single rod.

The molecular structure for a rod outer segment is schematized in Figure 5.15. A small area is enlarged to show the molecular packing of rhodopsin in the lamellar membranes. In fact, this model may be very close to reality, for recent x-ray diffraction studies of the frog rod outer segments seem to support such a model (Blaurock and Wilkins, 1969).

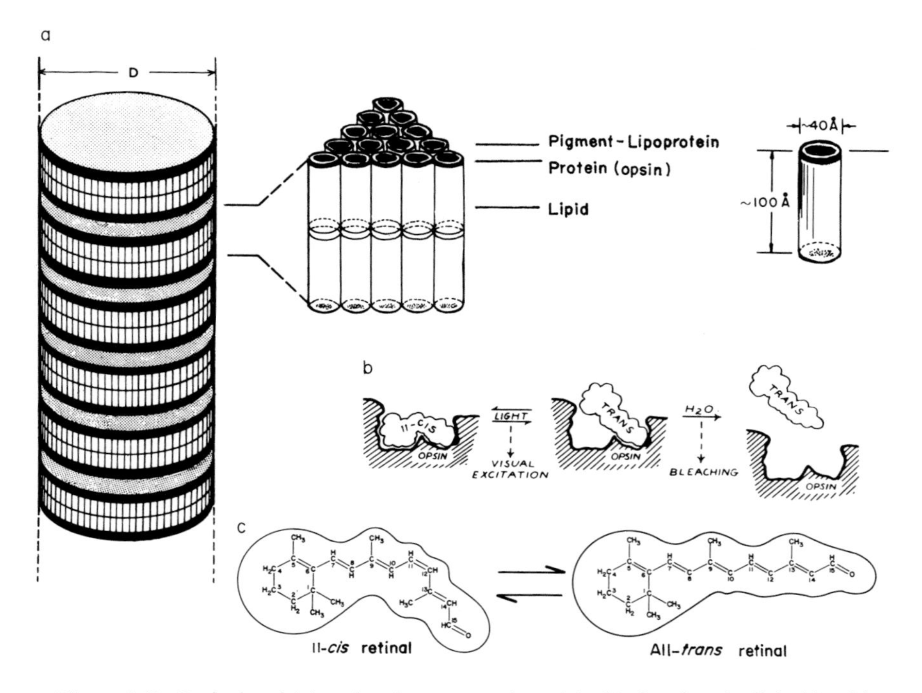

Figure 5.15. Retinal rod (a) molecular structural model; (b) showing the light bleaching of rhodopsin in the lamellae of the outer segment from the 11-*cis*- to the all-*trans*-retinal (from Hubbard and Kropf, 1959); (c) the molecular structure of 11-*cis*-retinal and all-*trans*-retinal.

The rhodopsin molecular weight M has also been calculated from the retinal rod structure. Where D is the diameter, T the thickness of the lamellar membranes, s the density (taken as 1.3 for a protein or 1.1 for a lipoprotein), L Avogadro's number, n the number of lamellar membranes, and N the number of rhodopsin molecules, the molecular weight is then obtained by

$$\mathrm{M} = \frac{\pi D^2 T s \mathrm{L} n}{4N}$$

The molecular weights calculated from this equation (density = 1.3) for frog and cattle rhodopsin were found to be 60,000 and 40,000, respectively (Table 5.2). This compares well with the molecular weight of 40,000 calculated by Hubbard (1954) for cattle rhodopsin. If, in calculating M, the density of a lipoprotein (1.1) is used, the molecular weight would be reduced by 20% resulting in a molecular weight of 32,000 for cattle rhodopsin and 54,000 for frog rhodopsin (Wolken, 1966). Heller (1969) has recently shown that the molecular weight of cattle, rat, and frog rhodopsin is about 28,000. However, the extraction procedures for the lipids of the visual complex and the extinction value Heller (1969) used for the calculation were questioned by Poincelot and Abrahamson (1970), and Shichi (1970).

Nevertheless, the highly ordered lamellar structure of the retinal rod outer segments is interpreted by us (Wolken, 1956b, 1961a) as a lipid–lipoprotein membrane (Figure 5.15). Such a structure provides sufficient space on its surfaces to accommodate all the rhodopsin molecules. This membrane structure maximizes the surface area available for light capture and permits the interaction of the chromophore retinal with both the lipid and protein. For, according to Kimbel *et al.* (1970), this transfer of the chromophore from the lipid to the protein upon illumination is the key molecular event in the generation of an electrical impulse.

Now that we have some basic information about the nature of vertebrate visual pigments, we must go back and ask several questions about the invertebrate photoreceptors. Are their visual pigments comparable to the vertebrate rod and cone pigments? Do we find in the arthropod photoreceptors (for example, in the honeybee, which exhibits color discrimination) visual pigments similar to those of the vertebrate cones? Also, are the invertebrate visual pigments associated in a very specific geometry with the membranes of their microtubules, as we have just hypothesized for the molecular structure of the vertebrate rod? Research attempting to answer some of these questions regarding the invertebrate visual pigments will be presented in Chapter VI.

VI. THE INVERTEBRATE EYE AND ITS VISUAL PIGMENTS

The Visual Pigments

We now come to grips with the question "What kind of visual pigments do the invertebrates possess?" The invertebrates present such a diverse group of animals and equally diverse visual photoreceptors, from eyespots, ocelli, and compound eyes to refracting eyes. Even though relatively few visual pigments have been isolated and identified we can piece together the experimental data from behavior, the extracted eye pigments, electrophysiology, and microspectrophotometry.

In the search to identify the visual pigments of invertebrates, it was anticipated that they would not be too different from the vertebrates. For that reason, we reviewed the vertebrate visual pigment chemistry in the preceding chapter. It will be remembered that all the identified vertebrate visual pigments are complexes in which the chromophore 11-*cis*-retinal is attached to a specific protein, opsin, to form a rhodopsin. We then described how rhodopsin was associated with the lipoprotein of the lamellar membranes of the rod outer segment (Figure 5.15). We noted that the most important step for visual excitation is that upon irradiation by light, rhodopsin liberates all-*trans*-retinal from opsin (Figures 5.12 and 5.13).

With this knowledge of the vertebrate visual system, we can now turn to the invertebrates to see if there is a universal pigment chemistry for all visual systems.

Molluscs

The cephalopod mollusc visual pigments were among the first to be isolated and studied. The extraction was accomplished with 1–2% aqueous digitonin, a method used for extracting vertebrate rhodopsin. The visual pigment obtained from the eyes of the squid, *Loligo pealeii,* with a maximum absorption at 493 nm was identified as a rhodopsin (Bliss, 1943, 1948; St. George and Wald, 1949). However, in contrast to the vertebrate rhodopsin system (Figure 5.13), the squid visual process only involves the transformation of rhodopsin to metarhodopsin. The transformation of metahodopsin to retinal and opsin does not occur except after extended exposure of the metarhodopsin to elevated temperatures. The regeneration of rhodopsin takes place directly from the metarhodopsin, and furthermore, it cannot accomplish this in the dark as with vertebrate rhodopsin, but rather in the light. In the light, then, a continuous equilibrium is set up between rhodopsin and metarhodopsin (Hubbard and St. George, 1958).

$$\text{rhodopsin (11-}cis\text{)} \overset{\text{light}}{\rightleftharpoons} \text{metarhodopsin (all-}trans\text{)}$$

Squid metarhodopsin can exist in two forms, depending upon the pH of the solution. In acid solution its maximum absorption is near 500 nm, whereas in alkaline solution, its maximum absorption is at 380 nm. These two forms are referred to as acid and alkaline metarhodopsin. They are interconvertible (in the dark only) simply by changing the pH of the solution. The reaction hinges on the addition of a single hydrogen ion to the alkaline metarhodopsin.

When the metarhodopsin is denatured with acid, the product, called "indicator yellow," is also pH-sensitive, having λ max about 440 nm in acid solution and 370 nm in alkaline solution. These are rather complicated reactions, but they explained the earlier observation of Krukenberg (1882) that squid rhodopsin either in the retina or in solution does not "bleach" on exposure to light. Bliss (1943, 1948) found that bleaching did occur when formaldehyde was added to his extracts. This is analogous to the denaturation of metarhodopsin by acid.

Since, in alkaline solution, metarhodopsin absorbs light at 380 nm, the composition of the rhodopsin–metarhodopsin equilibrium mixture can be shifted in either direction, depending upon the wavelength of light used. Near-ultraviolet light, which is absorbed more strongly by alkaline metarhodopsin, results in a larger quantity of rhodopsin in the equilibrium mixture, while orange light, absorbed more strongly by rhodopsin, produces a larger quantity of metarhodopsin. Irradiation with white light maintains the equilibrium, and irradiation of an acid solution produces a similar mixture, since both components absorb light in the same region of the spectrum.

The isomers of retinal involved in the visual process are all-*trans*-retinal, the chromophore of metarhodopsin, and 11-*cis*-retinal, the chromophore of rhodopsin. It has been suggested that the squid opsin molecule can accommodate both the all-*trans* and 11-*cis* forms of retinal, and therefore free retinal is produced only when the system is exposed to conditions that would denature opsin, and not as part of the system's normal functioning. That is, metarhodopsin—the stable combination of opsin and all-*trans*-retinal — is produced, and upon further irradiation, the retinal is isomerized to the 11-*cis* form, producing rhodopsin. As in the vertebrates, this interconversion is blocked by low temperatures (−65°C), which prohibit the necessary rearrangement of the opsin molecule.

Visual pigments have also been extracted in the same manner from the cuttlefish, *Sepia officinalis*, with a maximum absorption at 492 nm and from the octopus, *Octopus vulgaris*, with a maximum absorption at 475 nm (Hubbard and St. George, 1958; Brown and Brown, 1958). This is illustrated in the spectra of Figure 6.1a: octopus rhodopsin has a maximum absorption at 475 nm (curve 1); upon irradiation, metarhodopsin is formed (curve 2). Metarhodopsin, being a pH indicator, absorbs maximally in alkaline solution at 380 nm (curve 3), and at 503 nm in acid solution (curve 2). When metarhodopsin is denatured by acid, pH 3, it forms "indicator yellow" (Figure 6.1b) with a maximum absorption at 443 nm (curve 1), in neutral solution a mixture of free retinal (λ max 385 nm) and denatured opsin (curve 2), and in alkaline solution (curve 3) a mixture of retinal and opsin (Brown and Brown, 1958).

Therefore, in the photochemistry of the cephalopod rhodopsin, free

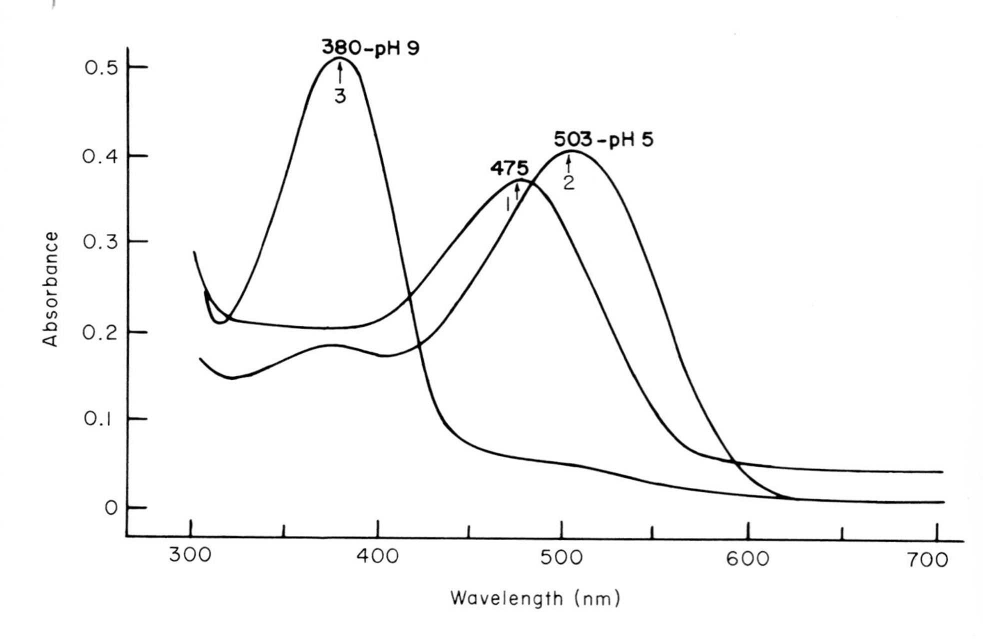

Figure 6.1a. Invertebrate rhodopsin. *Octopus*. Curve 1, rhodopsin, λ max 475 nm; curve 2, metarhodopsin, formed by irradiation, λ max 503 nm, pH 5; curve 3, metarhodopsin, λ max 380 nm, pH 9.

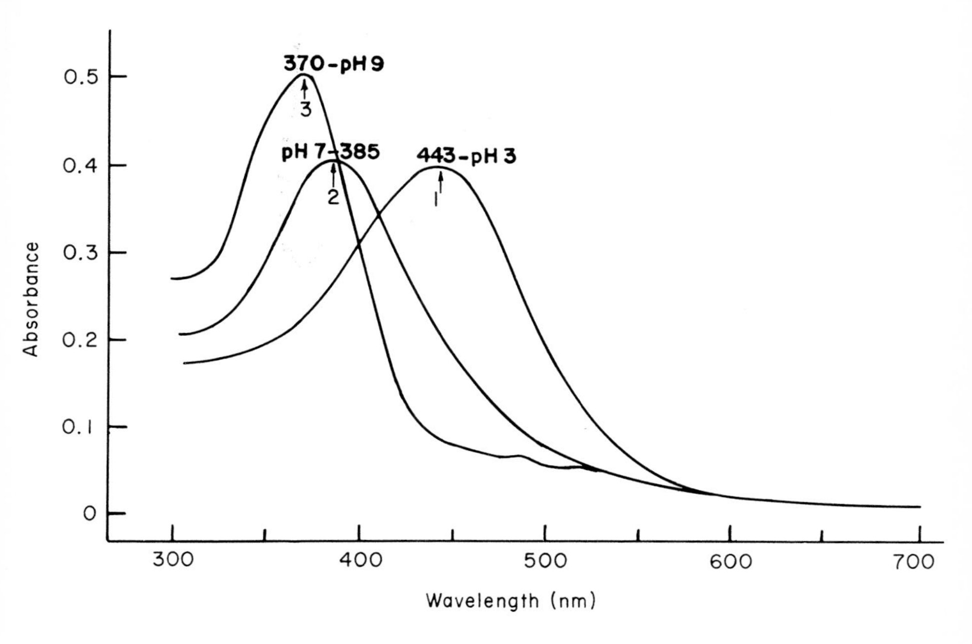

Figure 6.1b. Indicator yellow, formed by denaturating metarhodopsin, λ max 443 nm in acid, pH 3 (curve 1), and a mixture of free retinal, λ max 385 nm, pH 7.0 (curve 2); in alkaline solution, pH 9, λ max 370 nm (curve 3). (Spectra from Brown and Brown, 1958).

retinal is not liberated in the process. However, these mollusc visual pigments exhibit a basic similarity to vertebrate rhodopsins in that 11-*cis*-retinal is complexed with a protein opsin to form rhodopsin.

Arthropods

Crustacea

Among the crustacea, a visual pigment was extracted from the lobster *Homarus americanus* whose maximum absorption was about 515 nm (Wald and Hubbard, 1957). This value was in agreement with the spectral sensitivity as determined from its electroretinogram (ERG). On irradiation, the lobster visual pigment yields a stable metarhodopsin, 490 nm, whose behavior is similar to that described for the cephalopod metarhodopsins. Here too, retinal_1 is present in the 11-*cis* isomeric form. In addition, considerable vitamin A was found, but only in the lobster eyes and not in their bodies.

The visual pigment of the euphausiid shrimp, *Euphausia pacifica* was

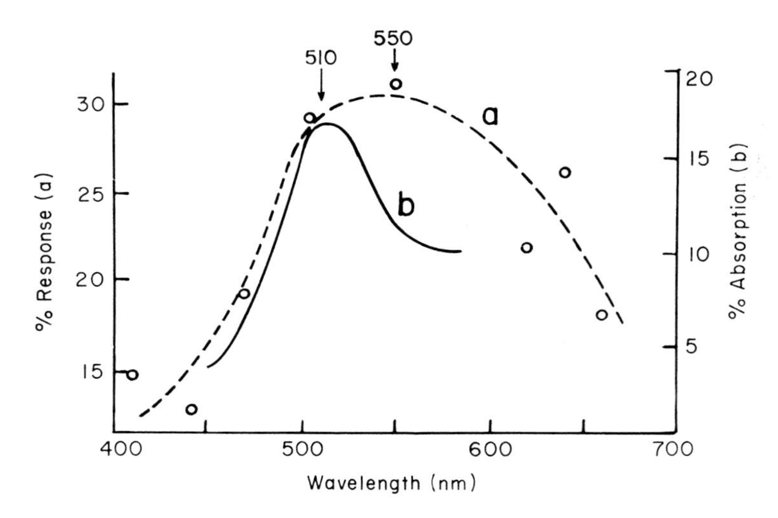

Figure 6.2. (a) Action spectrum of *Leptodora kindtii,* compared to (b) absorption spectrum of the rhabdom area obtained with microspectrophotometer.

found to have maximal absorption at 462 nm (Kampa, 1955), although some doubt was expressed about the accuracy of this value (Dartnall, 1962). In the euphausiid, *Meganyctiphanes norvegica,* the absorption maximum was also found to be near 462 nm. Here, retinal$_1$ was found to complex with opsin in the 11-*cis* configuration to form rhodopsin, and vitamin A was also found to be present (Wald and Brown, 1957).

In the northern crayfish, *Procambarus,* the maximum spectral sensitivity was found to be near 570 nm. This is further toward the red part of the spectrum than any of the known invertebrate rhodopsins (Kennedy and Bruno, 1961). Wald (1968) also extracted from the crayfish, *Orconectes virilis,* two photosensitive pigments with absorption maxima at 510 and 562 nm. The maximum near 510 nm lies close to the value for vertebrate rhodopsin, and the peak near 562 nm is close to the vertebrate cone pigment iodopsin (Figure 5.7). Wald suggested that these two pigments with different absorption maxima could function for color vision in the animal.

The spider crab, *Libinia emarginata,* as in the crayfish, has two visual pigments differentiated not for color vision, but for use in dim light and bright light, as are the vertebrate rods and cones (Wald, 1968). Microspectrophotometry of rhabdoms revealed only one pigment with a maximum absorption at 493 nm.

In the common prawn, *Palaemonetes vulgaris,* two different visual receptors seem to be present with maximal sensitivities in the red at

540 nm and in the violet at 390 nm (Wald and Seldin, 1968). However, microspectrophotometry of the rhabdom indicates two peaks in the visible near 496 and 555 nm (Goldsmith *et al.,* 1968).

For *Leptodora* (Figures 4.6–4.8), studied in our laboratory, the behavioral action spectrum indicated a broad band of absorption maximal around 540 nm (Figure 6.2a), and microspectrophotometry of the rhabdom yielded a narrow spectrum with a peak near 510 nm (Figure 6.2b). Here, as in the other crustaceans, these spectra may indicate that two distinctly different spectrally absorbing visual pigments are involved. However, pigments which exhibit these absorption peaks have not been successfully extracted.

Much research on the visual pigments of crustacea needs to be done in order to relate and clarify the photobehavioral action spectra, the microspectrophotometry of the rhabdom, the extracted pigments, and the electrophysiology for spectral sensitivity.

Insects

We indicated the great diversity in behavior found among insects and described their photoreceptor structures in Chapter III. Many insects are diurnal, well-adapted for navigating in bright light, and are able to distinguish colors; on the other hand, many insects are nocturnal and color blind (Wigglesworth, 1964). However, until recently very little was known about insect visual pigments (Wigglesworth, 1949; Wulff, 1956).

This is understandable, for to extract and isolate the visual pigments from insect eyes is extremely difficult; it requires that thousands of insects be collected, dark-adapted, decapitated, and their eyes dissected, all in the cold and in red light. Thus, to avoid these tedious procedures, researchers have preferred to gather most of their information from behavioral studies in which the spectral sensitivity, or action spectrum, could be related to the absorption of their visual pigments. Electrophysiology has been used as a tool to obtain the spectral sensitivity from electroretinograms (ERG). From behavioral and electrophysiological data, two receptor pigments have been inferred, one in the visible with an absorption maximum around 500 nm and another in the ultraviolet maximal near 365 nm. The absorption maximum around 500 nm suggests a rhodopsin visual pigment.

Drosophila. In this survey of insects, I would like to begin with *Drosophila melanogaster,* whose various eye-color mutants have been of great importance in biochemical genetics and whose eye structure we have described (Figures 3.7 and 3.8). The eye-color pigments are contained in four types of cells: (1) the preretinal primary cells found in the crystalline cone, (2) the secondary cells which surround the retinula cells, (3) the basal cells located at the basement membrane, and the (4) postretinal cells in the outer optic ganglion (Nolte, 1950).

It was suggested that a link between the eye pigmentation and neural function could be the tyrosine pathway common to the synthesis of melanins and catecholamines, which are used as chemical transmitters in various synapses (Hotta and Benzer, 1969). The brown pigments are derived from tryptophan by way of kynurenine and 3-hydroxykynurenine (Kikkawa, 1941; Butenandt, 1952). It was postulated that the tryptophan-derived pigments are composed of metallic salt complexes, that the pigments are produced by the function of an enzyme having a specific metal as part of its prosthetic group, and that the enzyme function is controlled by a particular gene (Kikkawa *et al.*, 1955).

In three different *Drosophila* eye-color mutants, scarlet (st), white (w), and wild-type red eyes (Canton), it was shown, using the extraction method of Kikkawa *et al.* (1955), that the scarlet eye pigmentation is influenced by an iron and/or molybdenum-bearing complex and that the white eye is controlled by a nickel complex (Wolken *et al.*, 1957b). The pigments from the wild-type red eyes were found to be pteridines or pteridine derivatives (Forrest and Mitchell, 1954a,b). These pigments have not been implicated in the visual pigment chemistry but will be discussed later in considering the insect screening pigments.

Behavioral studies of the response an organism makes to light intensity and wavelength should give an action spectrum indicative of the absorbing visual pigments. Fingerman (1952) and Fingerman and Brown (1952, 1953) have demonstrated that *Drosophila* possesses color vision at high light intensities, but at low light intensities there is a "Purkinje shift" from photopic to scotopic vision, similar to the vertebrate shift from cone to rod vision. This shift suggests that there are two kinds of receptors or spectrally different absorbing pigments in the eye of *Drosophila*. A similar finding has been noted for the spider crab by Wald (1968). The phototactic response curves obtained by Fingerman and Brown (1952) indicated that the basic photosensitivity curve for *Drosophila* is that of the white-eyed mutant and that the other response curves differ from this only because of the screening effects of their eye-color pigments.

For the three eye-color mutants, scarlet, white, and wild-type red-eyed *Drosophila,* we found that the action spectrum shows a single maximum near 508 nm, as shown in Figure 6.3 (Wolken *et al.*, 1957b). This finding, if it derives from absorption by the visual pigment, is close to the maxima reported for vertebrate rhodopsin (Figure 5.7), and implies that *Drosophila* should have a rhodopsin-like visual pigment.

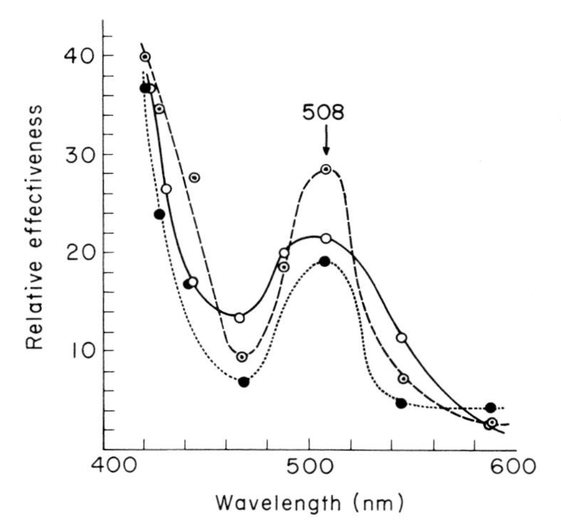

Figure 6.3. Action spectrum of *Drosophila melanogaster* for three eye-color mutants. ——o—— scarlet, – –⊙– – wild-type red, and ···●··· white.

To determine if this is so, the eyes of those three eye-color mutants were extraced with 1.8% aqueous digitonin, using the procedure for extraction of rhodopsin from vertebrate retinas. Spectra of these extracts showed a single maximum in the visible near 482 nm for the scarlet and wild-type red mutants, and for all three mutants a maximum in the ultraviolet at 342 nm and strong absorption in the region of 260–290 nm (Wolken *et al.,* 1957b). Even though these spectra would suggest a carotenoid pigment complex, no chemical identification for a carotenoid, vitamin A, or retinal was obtained. Also, it did not bleach like a visual pigment. Previously, Wald and Allen (1946) isolated from *Drosophila* a pigment that had a maximum at 436 nm, which was not photosensitive, and no positive tests for vitamin A or retinal were obtained. Later, using larger numbers of flies, Wald and Burg (1957) still were unable to identify for certain whether vitamin A was present. Most probably these extracted pigments are related to the ommochrome and pteridine screening pigments and are not visual pigments (Figure 6.19). Therefore, even in *Drosophila,* the most actively studied of the insects, the isolation and identification of a rhodopsin-like visual pigment remains inconclusive.

The Honeybee. Determination of the visual pigment of the honeybee, *Apis mellifera,* has been of great interest since von Frisch (1914) demonstrated that it could be trained to distinguish red, yellow, and green from blue and violet. Kühn (1927) showed that blue-green and near ultraviolet could also be distinguished by the bee. These behavioral studies were confirmed and extended by Bertholf (1931), Hertz (1939), Daumer (1956), and Kuwabara (1957), who concluded that the primary ranges of bee sensitivity are the near-ultraviolet 300–400 nm, blue 400–500 nm, and yellow 500–600 nm. Using electrophysiological methods, Goldsmith (1958a,b; 1960) showed that there should be several different absorbing visual pigments, one near 440 nm, another near 535 nm, and an ultraviolet sensitivity near 345 nm. Autrum and von Zwehl (1964) demonstrated that there were three different receptors for the worker bee with spectral sensitivity peaks at 340, 430, and 530 nm (Figure 6.4).

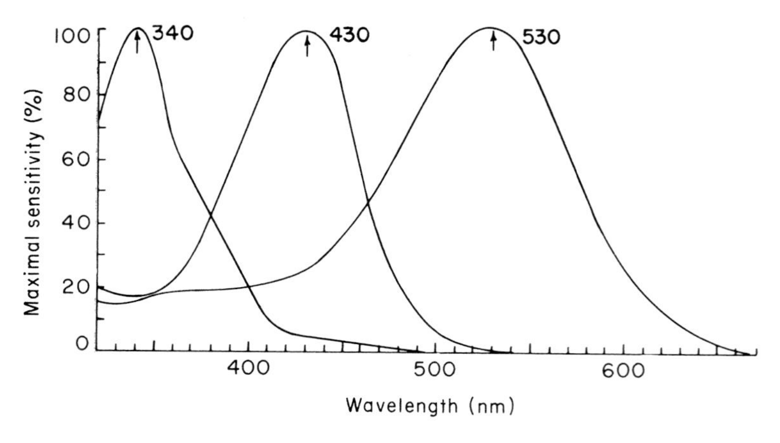

Figure 6.4. Spectral sensitivities for the worker bee *Apis mellifera* (from Autrum and von Zwehl, 1964).

To obtain the visual pigment from honeybees, the insects were previously dark-adapted, decapitated, and after grinding, extracted with phosphate buffer, pH 7. The extract was further purified and a pigment isolated which was found by difference spectra (obtained by several minutes exposure to a yellow light) to have a maximum near 440 nm and a negative maximum near 370 nm (Figure 6.5). This latter peak was attributed by Goldsmith (1958a,b) to the formation of retinal$_1$, for it was one of the peaks he found in the electrophysiological measurements.

Goldsmith (1958a) was able to show that retinal$_1$, presumably the 11-*cis*-retinal$_1$, was indeed present in the honeybee heads (Figure 6.6). He estimated that there was a maximum of 0.22 μg of retinal per gram of fresh heads, which corresponded to 3×10^{-6} μmoles per eye. No retinal could be extracted from the bodies. Furthermore, unlike all other visual pigments so far isolated, that of the bee could be brought into solution

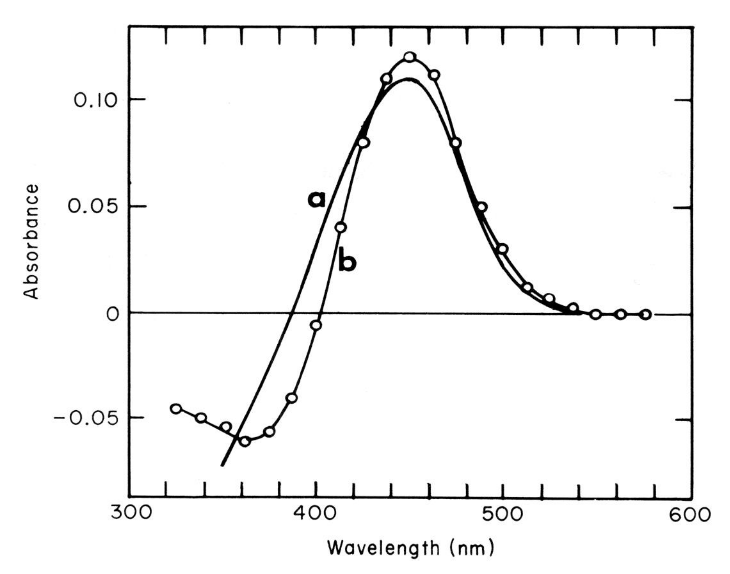

Figure 6.5. The honeybee, *Apis mellifera,* spectrum (a) difference spectrum of extracted photosensitive pigment, compared to spectrum (b) obtained by Goldsmith (1958b).

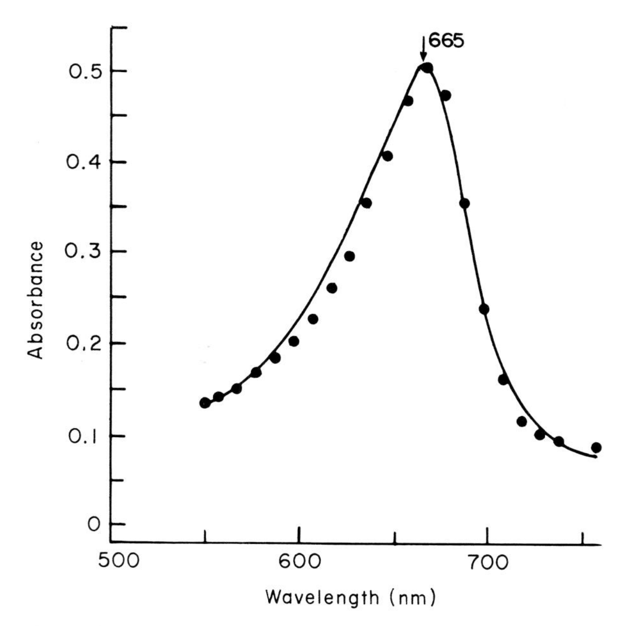

Figure 6.6. Absorption spectrum of antimony trichloride reaction of retinal isolated from dark-adapted bee heads (————), compared to absorption spectrum of crystalline retinal$_1$ (indicated by the points). This fraction of the head extract was eluted from alumina with 1% acetone in petroleum ether (Goldsmith and Warner, 1964, p. 435).

by grinding the heads of dark-adapted bees with ten times their weight of neutral phosphate buffer.

The finding that retinal$_1$ occurred in the head of the honeybee raised many questions about other insect visual pigments. Retinal is formed in the vertebrate eye by the oxidation of vitamin A, which is apparently not required by insects, for there has been no adequate evidence that vitamin A occurs in the bodies of insects. Goldsmith and Warner (1964) showed that vitamin A_1 is confined only to the honeybee *(Apis mellifera)* heads, that the vitamin A is converted to retinal on dark-adaptation, and that during light-adaptation, the amount of vitamin A increases.

The Housefly. We have pursued this same problem of visual pigments in the housefly, *Musca domestica,* and from their heads a light-sensitive yellow pigment was extracted with phosphate buffer. The yellow pigment was separated from a number of other pigments by chromatography on a column of calcium phosphate mixed with celite (Bowness and Wolken, 1959). This chromatographic procedure was based on the method used for purification of cattle rhodopsin (Bowness, 1959). The isolated pigment appeared to have many of the properties of the honeybee pigment, although its absorption maximum was shifted slightly down from 440 nm to 437 nm.

In obtaining the pigments from the housefly heads the following operations were carried out. The heads were extracted with 0.05 *M* phosphate buffer, pH 7.0, and centrifuged; the resulting supernatant was a dark red-brown color. When this was placed on the column, a light yellow fluid drained through. The presence of three other pigments became apparent on eluting with 0.2 *M* phosphate buffer, pH 6.5. A pigment with maximal absorption at about 545 nm and another maximal at about 408 nm were evident in the effluent fractions. The fourth pigment, the last to come off the column with 0.2 *M* phosphate buffer, was yellow and light-sensitive, and at pH 6.5 had an absorption peak at 437 nm (Figure 6.7). This last fraction containing the 437 nm pigment showed no trace of the absorption maxima of the other three pigments eluted earlier from the column.

After all the light-sensitive pigment had been eluted, a dark red-brown material, which could be eluted with 2 *N* KOH, remained at the top of the column. If the pigment was allowed to stay on the column for a day, it turned red, and immediately after elution with 1 *M* acetate–acetic acid buffer, pH 4.8, showed an absorption peak at 490 nm that changed to 440 nm on standing (Figure 6.8).

When the photosensitive yellow pigment was bleached by light at pH 6.5 or in the dark at pH 8.0, there was a shift toward longer wavelengths in the 500–540 nm region. In the light at pH 6.5, the main absorption

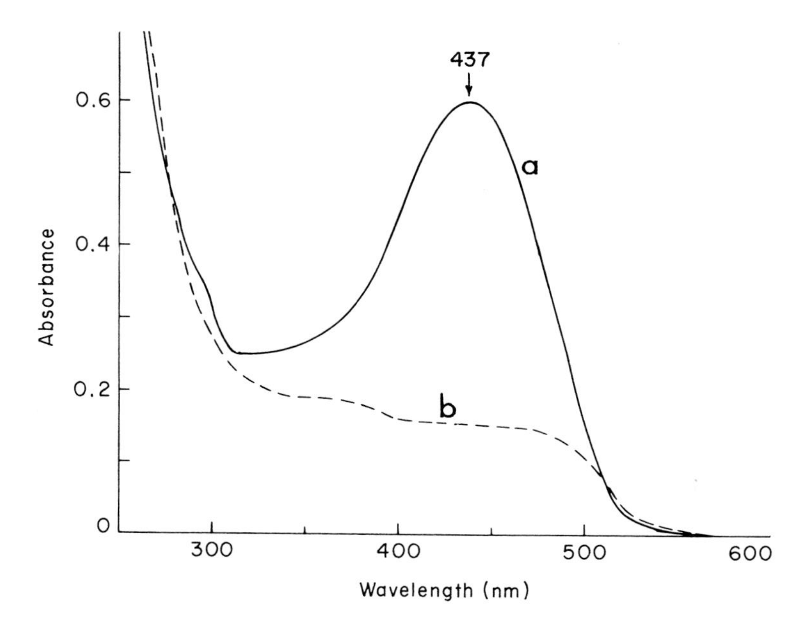

Figure 6.7. Absorption spectra of the housefly, *Musca domestica* photosensitive eye pigment. (a) Eluate P with 0.2 *M* phosphate, pH 6.5. (b) After bleaching with white light 600 fc for 2 hours at 12°C.

peak was gradually replaced by a plateau of absorption around 440–460 nm. In dilute solutions at pH 8.0, peaks at 250 and 290 nm were very pronounced. Difference spectrum curves showed maxima at 437 nm (Figure 6.9).

It is of interest to note a number of similarities in spectroscopic properties between this insect pigment and the vertebrate visual pigments reported by Wald (1955, 1959) and Dartnall (1957). First, there are the pH indicator properties shown by the light-sensitive housefly pigment and by its bleached products. On bleaching in the light at pH 6.5, a solution with plateaus of absorption at 440–460 nm and 350–360 nm is produced. Addition of a strong acid to this solution shifts the absorption peak close to 470–475 nm. In alkaline solution there is a plateau at 360 nm only. Absorption maxima in these three wavelength regions are given by the retinylideneamines and indicator-yellow under similar pH conditions (Ball *et al.,* 1949; Collins, 1954), although the 440 nm form of retinylideneamines is not stable except at pH 1 (Morton and Pitt, 1955). At pH 12 the housefly pigment produces an absorption maximum at 380 nm. A peak at this wavelength is obtained with Squid metarhodopsin at

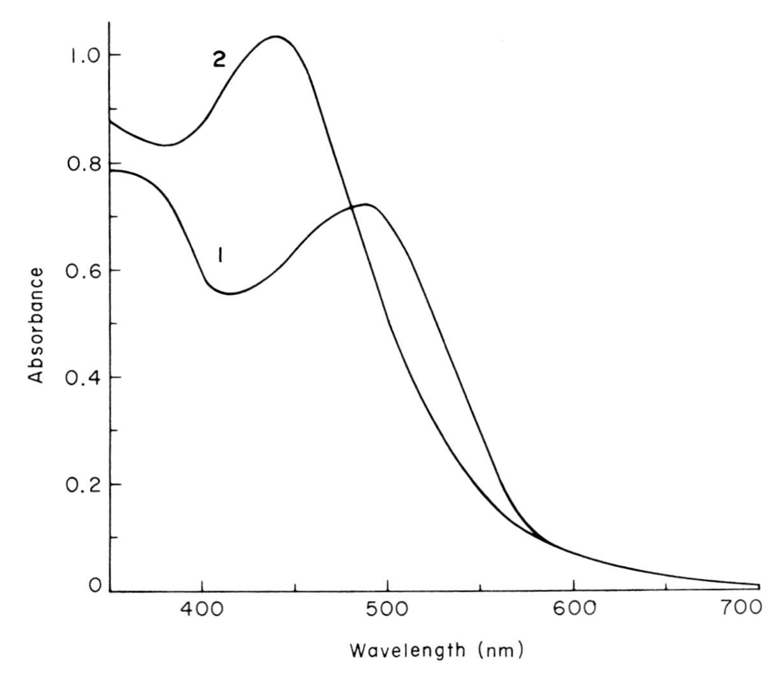

Figure 6.8. Pigment extract from housefly, *Musca domestica*. Curve 1, spectrum of extract made with 1 *M* acetate–acetic acid buffer, pH 4.8, of the top 2 cm of the extruded calcium phosphate and celite column, the loaded column having been eluted with 15 ml of phosphate buffer at pH 7.0 and 50 ml at pH 6.5. Curve 2, spectrum of pigment extract (curve 1) allowed to remain in the dark for 2 hours.

pH 9.9, and from cattle metarhodopsin at pH 13 (Hubbard and Kropf, 1959; Hubbard and St. George, 1958). Secondly, the bleaching process of the housefly pigment, as with vertebrate visual pigments (Wald, 1955), appears to involve more than one stage (Figure 5.13). Thirdly, the ultraviolet absorption spectrum of the pigment, either bleached or unbleached, has a peak at 290 nm in alkaline solution. Protein may be part of this light-sensitive housefly pigment, for most proteins show a peak at 290 nm in alkaline solution, though in neutral or acid solution they have a similar peak at 275–280 nm (Beaven and Holiday, 1952). Also, the heat bleaching of the pigment at 100°C gave a coagular precipitate which when dissolved in 0.2 *N* sodium hydroxide, showed a peak at about 290 nm. Fourthly, a precipitate containing about 10.5% nitrogen was obtained from the light-sensitive pigment solution upon addition of sulfosalicylic acid.

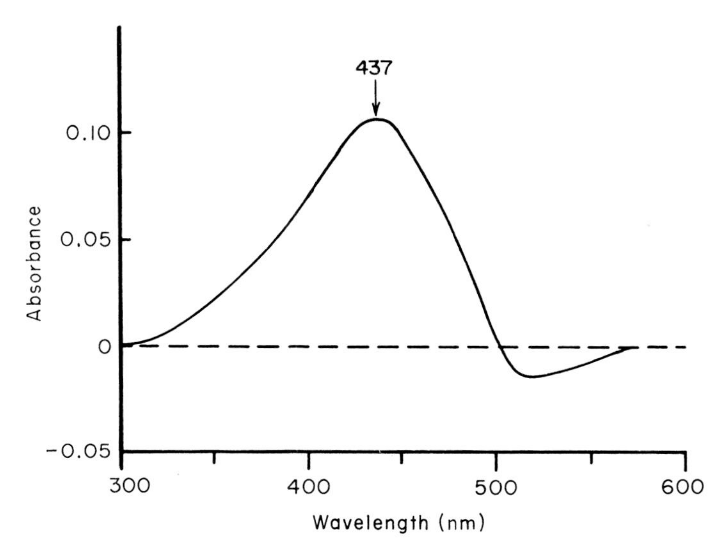

Figure 6.9. Housefly, *Musca domestica,* difference spectrum obtained from the eluate P of Figure 6.7, after bleaching for 10 minutes at 12°C with white light at 200 fc.

Of the other pigments, the light yellow pigment that drained through the column with 0.025 *M* phosphate buffer appears to be of the melanin type. The red pigment, which required 2 *N* KOH or 1 *M* acetate-acetic acid buffer, pH 4.8 for elution from the column, exhibited a shift in absorption maximum from about 490 to 440 nm, in changing from alkaline to acid conditions (Figure 6.8). This is similar to the shift shown by rhodommatin, a red ommochrome-type pigment (refer to Figure 6.19) obtained from insects by Butenandt *et al.* (1954).

To see whether another photosensitive pigment could be extracted whose absorption peak would be closer to the behavioral action spectrum, aqueous digitonin (1.8%) extraction and the purification procedures described for cattle rhodopsin were used (Bowness, 1959). The flies were dark-adapted prior to dissection and their heads ground and extracted. All operations were carried out in the dark and in the cold. By this method a pigment was eluted from the chromatographic column whose difference spectrum had an absorption maximum near 510 nm (Figure 6.10a). This absorption peak is closer to the spectral sensitivity peak for the housefly than the phosphate buffer-eluted photosensitive yellow pigment (Figure 6.7). In addition, we were able to confirm this absorption peak by microspectrophotometry in the rhabdom area of dark-adapted housefly eyes.

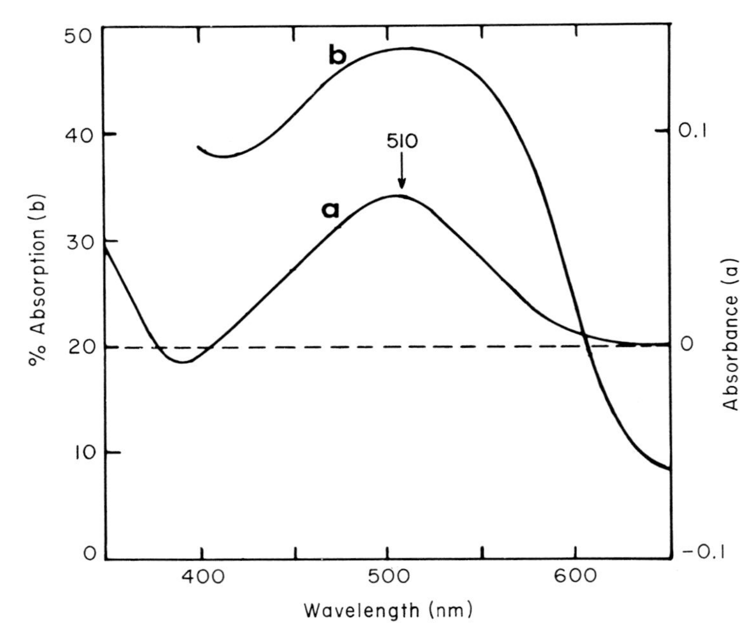

Figure 6.10. Housefly, *Musca domestica.* (a) Difference spectrum of digitonin extract of housefly eye pigment, λ max 510 nm, compared to (b) absorption spectrum of the rhabdom area, obtained by microspectrophotometry.

The absorption spectrum obtained closely approximates the digitonin-isolated pigment with a maximum absorption near 510 nm (Figure 6.10b).

Now, in order to identify whether the chromophore retinal is part of the visual pigment of the housefly eye, three different extracting procedures were used. First, the heads were completely ground with anhydrous sodium sulfate and acetone. The mixture was centrifuged at 3000 rpm and the supernatant poured off and retained. The residue was re-extracted with acetone, and after centrifugation it was combined with the first extract. This was repeated until all the color was removed.

The second method was to grind the heads with 80 ml 0.2 *M* phosphate buffer (pH 6.5) and to centrifuge this mixture at 12,000 rpm for 20 minutes. The residue was extracted with acetone, while the supernatant was fractionated by saturation with ammonium sulfate at 45% and 60%. The ammonium sulfate precipitates were then extracted with acetone.

The third method was to grind the heads with 20 ml of 50% sucrose in 0.06 *M* phosphate buffer (pH 6.5) and then to centrifuge the sucrose mixture at 3000 rpm for 10 minutes. The residue was extracted with acetone, while the supernatant suspension was diluted with phosphate buffer to a sucrose concentration of 12.5% and centrifuged at 14,000 rpm for 20 minutes. The sedimented material was then extracted with acetone.

The initial acetone extract from each of the above preparations was evaporated to dryness, and the residue was dissolved in petroleum ether. The petroleum ether extract was then dried over anhydrous sodium sulfate, redissolved in petroleum ether and chromatographed on alumina. The first column was eluted with 40 ml of 40% acetone in petroleum ether (v/v). The eluate was evaporated to dryness, redissolved in petroleum ether, and applied to a second column; fractions were eluted with 4%, 17.5%, 25%, and 40% acetone in petroleum ether (v/v). The effluent fractions were evaporated to dryness, then stored in a vacuum dessicator in the dark. The acetone extracts of the precipitate from the ammonium sulfate fractionation (procedure 2) and the extract of the material which sedimented in 12.5% sucrose solution (procedure 3) were evaporated to dryness and dissolved in chloroform.

The results of these chromatograms showed a separation of one or more of the carotenoids present. Of the various carotenoids obtained, $retinal_1$ was identified. The fraction eluted with 25% acetone in petroleum ether had an absorption maximum at 385 nm in chloroform (Figure 6.11), and when reacted with antimony trichloride, a transient blue color was observed, maximal at 665 nm (Figures 6.6 and 6.12). This was a positive test for the presence of $retinal_1$.

The presence of $retinal_1$ in the extract indicates that vitamin A_1 should also be present, since in the visual cycle $retinal_1$ is reduced to vitamin A_1. Vitamin A_1 has a principal absorption maximum at 328 nm in chloroform and it yields a blue color with the antimony trichloride reaction giving a peak at 620 nm. Fractions were obtained whose antimony trichloride tests gave maxima near 620 and 696 nm, the maxima for vitamins A_1 and A_2; in some cases these were found in the same fraction (Wolken *et al.*, 1960). The identification of vitamin A_2 is particularly crucial since it is so far known to occur only in freshwater fish and in some amphibians (Figure 5.8). This would be the first instance of its presence outside these vertebrate groups.

Of the other fractions eluted from the column, two contained light-stable carotenoids. The fraction eluted with 4% acetone in petroleum ether was yellow in color, and its absorption spectrum in chloroform showed maxima at 426, 455, and 486 nm, which correspond closely with those of lutein, a plant xanthophyll (Figure 1.6).

The amount of retinal in the housefly heads was estimated from the antimony trichloride reaction (Ball *et al.*, 1949), which gave us a value of

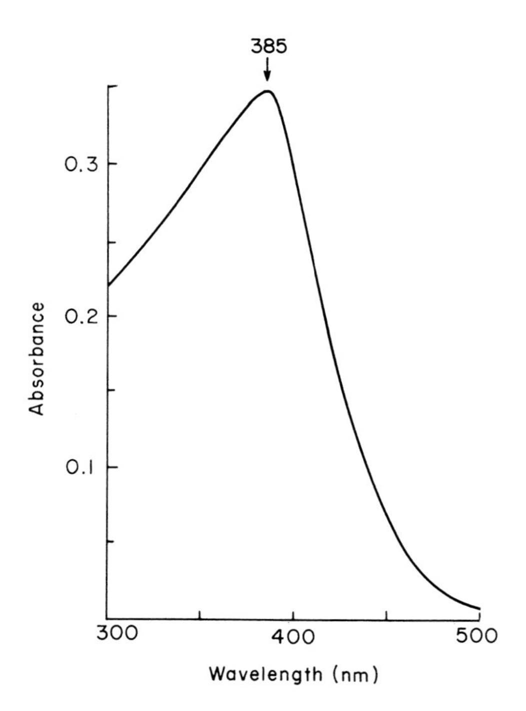

Figure 6.11. Housefly, *Musca domestica,* eye extract. Absorption spectrum of the retinal-containing fraction in chloroform.

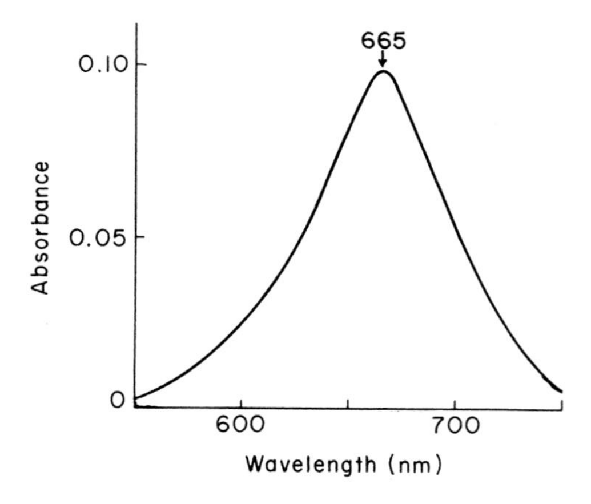

Figure 6.12. Housefly, *Musca domestica,* absorption spectrum of antimony trichloride reaction of the retinal-containing fraction immediately after mixing.

0.31 μg of retinal/gm weight of fresh heads. From the average number of 24,500 rhabdomeres per eye this corresponds to 3.7×10^7 molecules of retinal per rhabdomere (Wolken *et al.*, 1960).

The Cockroach. In our further search to identify the visual pigments of insects, let us turn to the cockroach, which is a nocturnal insect and is considered primitive in comparison with the honeybee and the housefly.

Electrophysiological studies of the cockroach eye (Walther, 1958; Walther and Dodt, 1959) show a maximum sensitivity peak at 507 nm. In addition, Walther (1958) suggested that the cockroach eye contains two types of photoreceptors; the lower half of the eye containing one type with an absorption maximum at 507 nm, and the upper half containing both types, one at 507 nm and another with maximal sensitivity between 314 and 369 nm.

Cockroach eyes are large in comparison with those of other insects and are therefore more easily dissected in the dark. To see whether their visual pigment could be spectrally identified *in situ,* sections were made through the rhabdom area of the eye of *Blaberus giganteus.* These sections were scanned with the microspectrophotometer which revealed an an absorption spectrum with a major peak around 495 nm (Figure 6.13), which is typical of a rhodopsin spectrum.

To determine chemically whether rhodopsin was the visual pigment, large numbers of eyes *(Blatta orientalis* and *Periplaneta americana)* were dissected and the eyes ground in 45% sucrose–phosphate buffer solution. This was followed by differential and high speed centrifugation to isolate a rhabdomere fraction containing the visual pigments. The

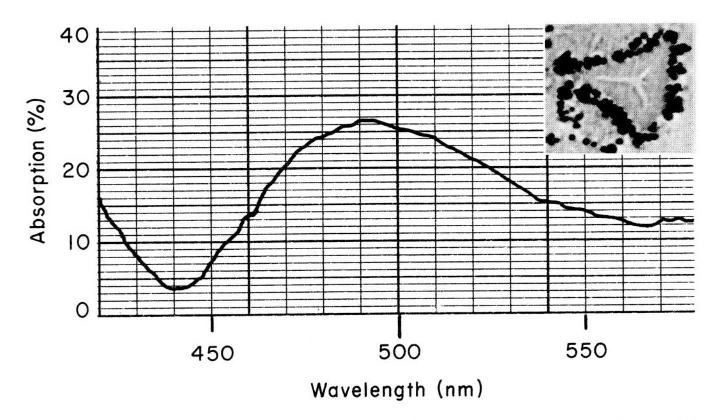

Figure 6.13. Cockroach, *Blaberus giganteus,* absorption spectrum obtained by microspectrophotometry of the rhabdom. Insert is a light micrograph of an osmium fixed and sectioned cockroach rhabdom.

rhabdomere fraction so obtained was then extracted with 1.8% aqueous digitonin. This resulted in the extraction of a photosensitive pigment, with a maximum absorption peak near 500 nm (Figure 6.14). The difference spectrum, of the light bleached against the unbleached pigment, showed maximum absorption about 500 nm corresponding to a rhodopsin, and a negative maximum at 375 nm corresponding to retinal (Figure 6.15). The light bleaching spectra showed a shift from the absorption peak at 500 nm to 375 nm, indicating the uncoupling of retinal from opsin in the rhodopsin complex.

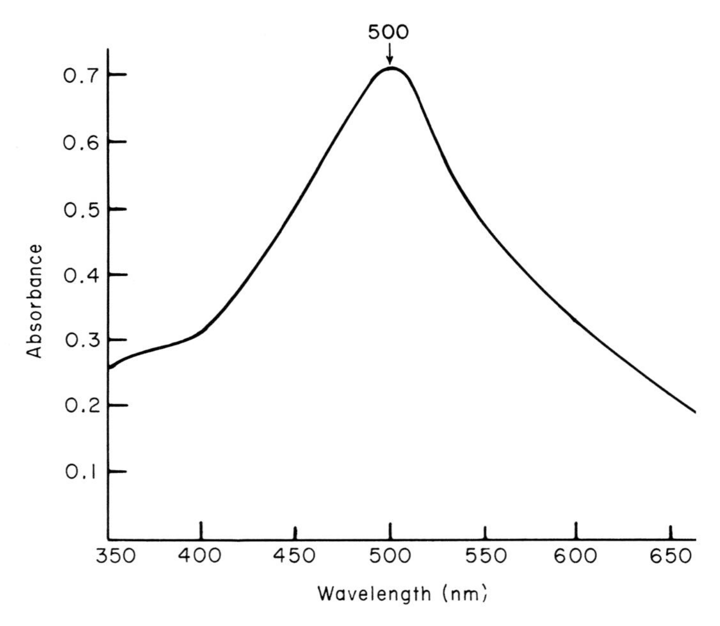

Figure 6.14. Cockroach, *Blatta orientalis,* absorption spectrum of the rhabdomere fraction extracted with 1.8% aqueous digitonin, pH 7.0, in 15 *M* phosphate buffer.

The time course of bleaching for the photosensitive pigment follows first-order kinetics, since the plot of the natural logarithm of the absorbency of the pigment complex against the time of exposure to light is linear. The first-order rate constant obtained from the slope of this line is 6.1×10^{-5} sec^{-1} at 25°C (Figure 6.16).

To determine whether retinal was really a part of this photosensitive pigment, extraction and column chromatography were employed (Wolken and Scheer, 1963). The chromatographic fractions eluted from the

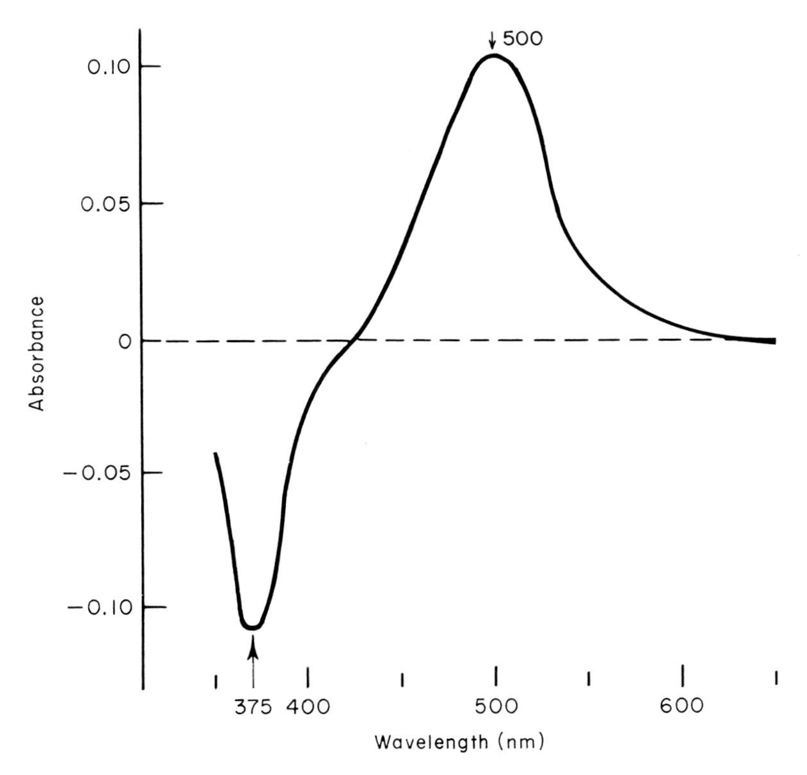

Figure 6.15. Difference spectrum of the cockroach *(Blatta orientalis)* eye pigment after exposure to white light (refer also to Figure 6.14).

alumina columns in acetone gave an absorption spectrum indicative of $retinal_1$. The fraction eluted with 17.5% acetone in petroleum ether, when reacted with antimony trichloride, showed a color change to blue and an absorption maximum at 664 nm, which is characteristic of $retinal_1$ (Figure 6.17).

The amount of retinal per gram of fresh cockroach eyes was estimated from the extinction coefficient of antimony trichloride to be 0.08 μg. On the basis of 28,000 rhabdomeres per eye (Wolken and Gupta, 1961), the amount of retinal or rhodopsin per rhabdomere was estimated to be 4.3×10^7 molecules. This compares favorably with the rhodopsin concentration for vertebrate rods, which contains from 10^6 to 10^9 rhodopsin molecules (Table 5.2).

We have succeeded in isolating only one photosensitive pigment from the compound eye of the cockroach (Wolken and Scheer, 1963). It is

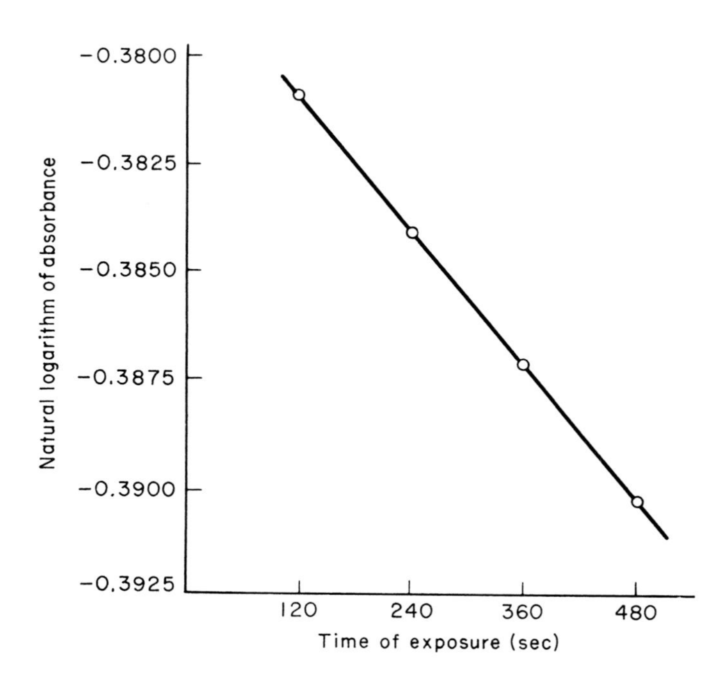

Figure 6.16. The cockroach *(Blatta orientalis)* eye pigment. Time course of bleaching of the photosensitive pigment.

possible that the structure and arrangement of the lens—which can play a significant role in determining the near ultraviolet sensitivity of the eye (Wald, 1945; Dodt and Walther, 1958; Buriain and Ziv, 1959)—together with the rhodopsin bleaching characteristics (Wald and Brown, 1958) could account for the dual sensitivity of the cockroach eye. This possibility becomes more interesting in view of the experiments of Miller and Bernard (1968) who have shown that the corneal lens fine structure permits it to function as an interference filter. On the other hand, Mote and Goldsmith (1970) found in the white eye mutant of *Periplaneta,* using intracellular recordings, two receptor cells: green receptor cells with a spectral sensitivity of about 507 nm, close to that of our isolated photosensitive pigment (Wolken and Scheer, 1963) and ultraviolet receptor cells with a maximum spectral sensitivity at 365 nm.

Spiders and King Crabs

The visual pigments of spiders have been little studied. It was thought for some time that spiders also see color (Peckman and Peckman, 1887). The most thorough study was done by Kästner (1950) on jumping spiders

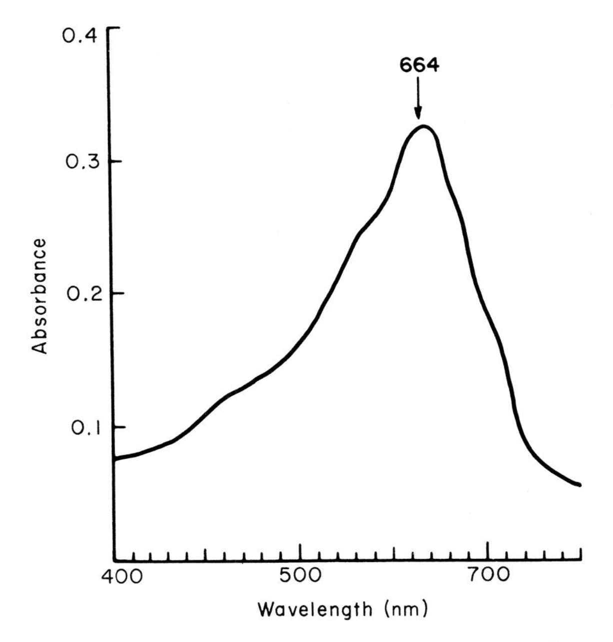

Figure 6.17. The cockroach *(Blatta orientalis)* eye pigment. Absorption spectrum of the fraction chromatographed on an alumina column, eluted with 17.5% acetone in petroleum ether and reacted with antimony trichloride ($SbCl_3$). The arrow indicates absorption maximum at 664 nm ($retinal_1$).

(Evarcha falcata), using a behavioral approach, with the conclusion that indeed these spiders had color vision. De Voe *et al.* (1969) showed that the wolf spider median eyes have a visual pigment absorbing maximally at 505–510 nm, but have a much greater sensitivity at 380 nm. This ultraviolet sensitivity is believed to be due to another pigment, not necessarily a primary visual pigment. From the horseshoe or king crab, *Limulus polyphemus,* a visual pigment with $retinal_1$ was extracted. This pigment absorbs maximally at 520 nm and bleaches on irradiation like vertebrate rhodopsin (Hubbard and Wald, 1960).

Accessory and Screening Pigments

Invertebrate eyes also possess screening and reflecting pigment granules which serve to regulate the amount of light that reaches the photoreceptors. What function these pigment granules may have in visual excitation is not completely known.

In vertebrates, which are dependent on bright light, the retinal cones predominate; for example, birds, turtles, lizards, and snakes. In these animals, situated only in the area between the inner and outer segment of the cones, there are oil globules pigmented red, green, yellow, and colorless (Figure 6.18). By differentially transmitting light to the outer segments, the oil globules can affect the spectral response of the cones and can thus provide the basis for color discrimination (Strother and Wolken, 1960b; Wolken, 1966).

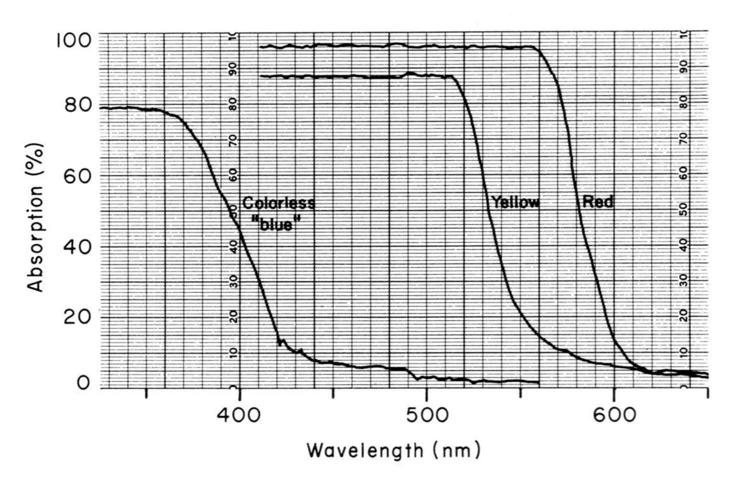

Figure 6.18. Absorption spectra of colored oil globules in the retina of the swamp turtle (*Pseudemys scripta elegans*).

It was thought that there would be similarities between these pigmented globules and the arthropod pigment granules. However, no histological or other kind of evidence has been found to indicate that they are chemically identical or have similar function. The pigmented oil globules are all carotenoids (Wald, 1948; Wolken, 1966), whereas the screening pigment granules, although they may differ in composition (Grossbach, 1957), have been primarily identified as ommochromes, pterines, and pteridines (Ziegler-Günder, 1956; Ziegler, 1964).

Ommachrome pigments are yellow to dark-red and fall into the class of ommatines and ommines (Linzen, 1959). The ommochromes of the xanthommatin type are photosensitive, they can be oxidized and reduced, and are pH sensitive (Figure 6.19). It was suggested (Yoshida *et al.*, 1967) that these ommochromes could function as do the quinones with

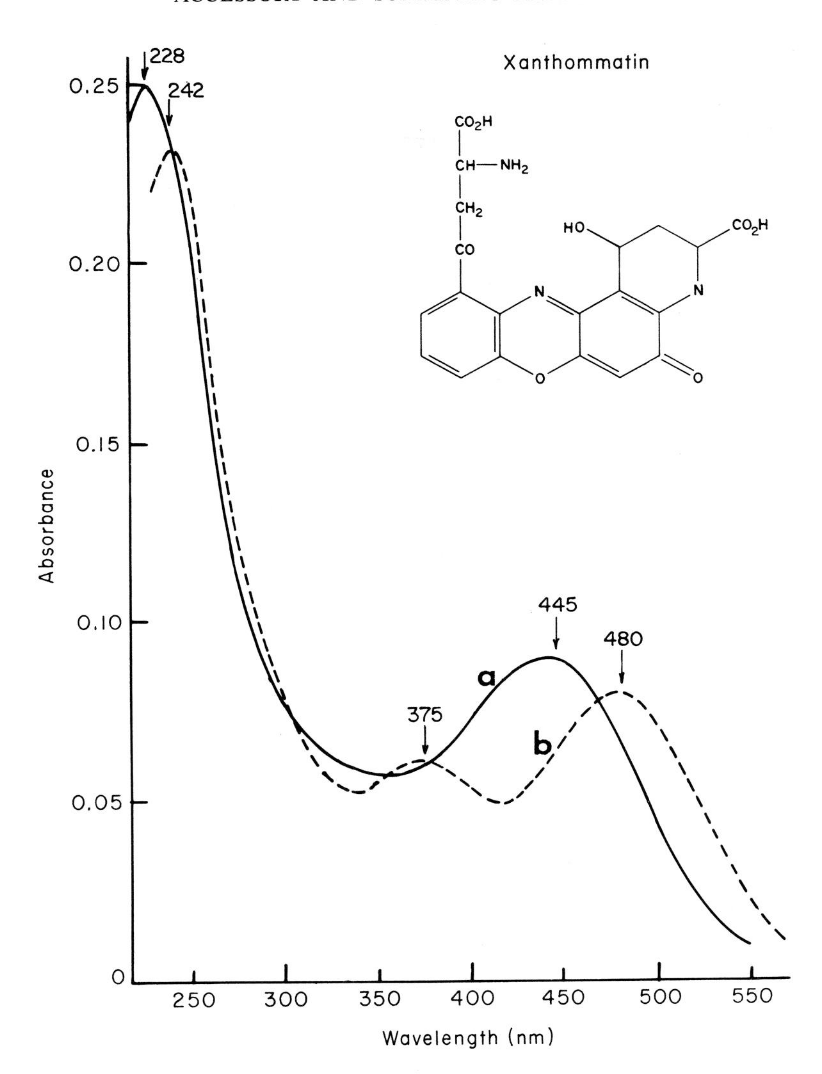

Figure 6.19. Absorption of spectrum of xanthommatin. Extracted from *Calliphora.* (a) pH 7 (———) and (b) in 5 *N* HCl, pH 2 (– – – –). (From Butenandt and Neubert, 1955, pp. 110–111.)

the cytochromes in the electron transfer chain (Figure 1.13). The pterine pigments are usually chemically represented as the yellow xanthopterin (2-amino-4-6-dihydroxypyrimidopyrazine ring) which upon ultraviolet excitation, fluoresces blue (Forrest and Mitchell, 1954a,b). The pterines are stable *in vivo* but photosensitive *in vitro*. Although they take part in some metabolic processes in the eye, they do not appear to be part of the primary photoreceptor molecule in visual excitation (Autrum and Langer, 1958).

In the housefly, *Musca domestica*, the yellow pigment granules are found at the top of the ommatidium close to the corneal lens and crystalline cone, whereas the red pigment granules are found at the bottom of the ommatidium and closer to the photoreceptor area, the rhabdom. Microspectrophotometry of the yellow and red granules *in situ* shows (Figure 6.20) that the yellow pigment in these granules has a broad absorption band maximal at about 440 nm, and that the red pigment also has a broad absorption band with its maximum nearer to 530 nm, with another smaller peak close to 390 nm (Strother and Casella, 1970).

The red pigment granule absorption spectrum (Figure 6.20a) shows that some light may be transmitted at 350 and 420 nm. However, the yellow pigment (Figure 6.20b), which is situated near the top of the eye, is highly absorbing from roughly 370 nm to 530 nm and therefore, it is unlikely that

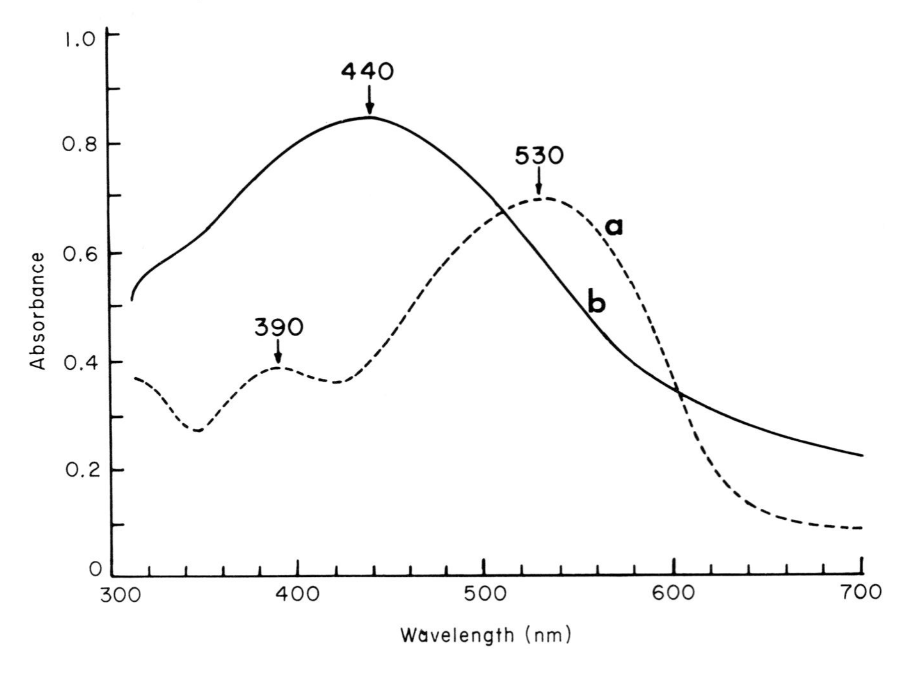

Figure 6.20. Absorption spectra of screening pigment granules of the housefly, *Musca domestica*, red granules (a) and yellow granules (b), Strother and Casella, 1970.

light leakage through the screening pigments is important in the region 370–620 nm for the housefly.

In the blowfly, *Calliphora erythrocephala,* Langer (1967) found that the yellow pigment granules have a single peak at about 445 nm and the red pigment granules showed peaks near 390 and 540 nm. The yellow pigment absorption spectrum resembles that of an oxidized xanthommatin (Figure 6.19a) and the red pigment that of a reduced xanthommatin–protein complex (Butenandt *et al.,* 1954; Butenandt and Neubert, 1955; Burkhardt, 1962).

These spectra are also interesting from a physiological standpoint, for no obvious correlation between them and the overall spectral sensitivity for these insects is evident. Accordingly, Langer and Thorell (1966) made direct microspectrophotometric measurements of the *Calliphora* rhabdomeres within a rhabdom. They found two different spectra; one had two absorption peaks with maxima at about 380 and 510 nm and closely resembled the red screening pigment spectrum (Figure 6.20a). The other showed only a single peak near 470 nm and its shape closely resembled the yellow pigment spectrum (Figure 6.19b). These results would indicate then that the screening pigments function to absorb light for the visual pigments. In fact, a photosensitive yellow pigment with an absorption peak near 440 nm was extracted from housefly eyes (Bowness and Wolken, 1959) and from the honeybee (Goldsmith, 1958a,b), which closely matches the yellow screening pigment absorption peak (Figures 6.19b and 6.20b). Data obtained by Langer (1967) and Strother and Casella (1970) indicate that for the blowfly and housefly the combined effect of both yellow and red screening pigments is to effectively screen the separate ommatidia from light leakage over a wavelength span of 320 nm to about 640 nm (Figure 6.20). However, it is possible that the yellow and red screening pigments are acting to separately screen two different visual pigments, namely the one absorbing at 440 nm and the other near 510 nm.

These spectra suggest that those regions where the pigments permit passage of light coincide with the spectral regions where the insect is most strongly sensitive. Thus ultraviolet light of 350–400 nm and red light beyond 600 nm are transmitted and not absorbed by the screening pigment granules. This transmitted light is then available to the photoreceptors and results in a greater sensitivity for the insect.

Concluding Remarks

The visual pigment of invertebrates is a rhodopsin in which the chromophore is retinal$_1$. Cephalopod and crustacean rhodopsin light bleaching involves only the transformation to metarhodopsin (Figure 6.1), that is, the liberation of free retinal and opsin does not occur as in the vertebrate photochemistry.

Spectral sensitivity measurements indicate that there are several absorbing pigments and that, like the rods and cones of vertebrate eyes, these are used for sensing bright and dim light and for color discrimination. For example, the lobster visual pigment shows absorption peaks at 515 and 480 nm, the crayfish at 562 and 510 nm, the honeybee at 430 and 530 nm, and the housefly at 437 and 510 nm (Table 6.1). In addition, there is strong absorption in the ultraviolet in the neighborhood of 340–390 nm.

TABLE 6.1
SOME INVERTEBRATE VISUAL PIGMENT, ABSORPTION MAXIMA

	Common name	Maxima (nm)
Molluscs		
Cephalopods		
Loligo peallii	Squid	493
Octopus vulgaris	*Octopus*	475
Sepia officinalis	Cuttlefish	492
Crustacea		
Homarus americanus	Lobster	515
Euphasia pacifica	Shrimp	462
Procambaras	Swamp crayfish	570
Orconectes virilis	Northern crayfish	510 and 562
Leptodora kindtii	water flea	510
Palaemonetes vulgaris	prawn	496 and 555
Insects		
Apis mellifera	honeybee	440
Musca domestica	housefly	437 and 510
Calliphora erythrocephala	blowfly	470 and 510
Blatta orientalis	cockroach	500
Periplaneta americana	cockroach	500
Blaberus giganteus	cockroach	495
Arachnid		
Limulus polyphemus	Horseshoe crab (king crab)	520

To clarify the dual spectral sensitivity and spectrally different absorption peaks of the visual pigment, Autrum and Burkhardt (1961) and Burkhardt (1962) measured spectral sensitivity for the blowfly, *Calliphora erythrocephala,* using microelectrodes placed in single retinula cells. They found three different spectral sensitivities in the visible with maxima about 470, 490, and 520 nm. Each was coupled with an ultraviolet band about 350 nm. In similar experiments with drone bees, Autrum and von Zwehl (1962) found two different receptors with maxima at about 340 and 447 nm. The 447 nm peak compares well with the observed spectral sensitivity maximum for drones at 430 nm (Autrum and von Zwehl, 1964) and to the extracted photosensitive honeybee pigment maximal near 440 nm (Goldsmith, 1958a,b).

In the blowfly *(Calliphora erythrocephala),* the eye-color mutant "chalky" lacks all eye screening pigments, and Langer and Thorell (1966) were able to obtain spectra of the rhabdom by microspectrophotometry. The *Calliphora* rhabdom consists of seven rhabdomeres of the open-type (see Figure 3.8); Langer and Thorell found that for six of the rhabdomeres the absorption maximum was about 510 nm and for the seventh or asymmetric rhabdomere the absorption maximum was about 470 nm. These spectral peaks come close to Burkhardt's maxima for spectral sensitivity, and presumably these absorption peaks are associated with two differently absorbing visual pigments.

To find that there are two differently absorbing pigments present in different photoreceptors raises the question of whether the arthropods have visual pigments which spectrally behave like the rods and cones of the vertebrate retina (Table 6.1; Figures 5.7 and 5.8).

With regard to the screening pigment granules which regulate the amount of light reaching the photoreceptors, they are for the most part ommochrome or pteridine pigments which are light- and pH-sensitive (Ziegler, 1965). All evidence indicates that the screening pigments function to absorb light for the visual pigment.

The corneal lenses are also of interest, for in the compound eye they can function as interference filters (Bernard and Miller, 1968; Miller *et al.,* 1968). In doing so, they could act, like pigment oil globules (Figure 6.18), to provide a system for color vision in these animals.

VII. SUMMARY AND CONCLUDING THOUGHTS

We began this comparative structural analysis of invertebrate photoreceptors by first considering the general phenomena of photosensitivity in nature. In doing so, we briefly reviewed the kinds and structure of the pigments responsible for this photosensitivity.

The molecular structure of these pigments revealed certain similarities which were noted in our discussion (Figures 1.3, 1.5, 1.6, 1.9, and 5.9). However, it was emphasized that the pigment, or chromophore, does not function alone but is complexed with a specific protein to form a photosensitive complex. Because these photopigments reside within cell organelles, we next had to consider both the kind and internal structure of these organelle photoreceptors.

What we were trying to find in these photoreceptor molecules and photoreceptor structures were molecular relationships common to all photoreceptors.

Our general thesis was that the key to understanding the photobehavior of invertebrates could be found in the pigment molecules and their molecular arrangement within the photoreceptor. To gain supporting evidence for this thesis, invertebrate visual systems and photoreceptors were examined in various invertebrate phyla to establish their molecular structure, identify their photosensitive pigments, and to see how they fit within the known scheme for the vertebrate visual system. In doing so, were there any evolutionary trends that could be discerned in the photoreceptor systems as we proceeded from the protozoa to the arthropods and to the molluscs? Let us return to summarize these findings.

Primitive Photoreceptors

In the protozoa, *Euglena* was taken as a model cell, for it possesses a primitive photoreceptor system consisting of an eyespot and a flagellum. *Euglena's* photobehavior is much like that of a photocell in which movement is a light-searching process.

The eyespot is an agglomeration of orange-red pigmented granules located around the gullet in which the flagellum is found (Figure 2.6).

Attached to the flagellum and facing the eyespot granules is the paraflagellar body, the photoreceptor (Figure 2.7).

The phototactic action spectra for *Euglena* (Figure 2.14) and absorption spectra of the eyespot (Figure 2.15a) indicate that the eyespot contains a carotenoid, e.g., β-carotene, but absorption spectra obtained by microspectrophotometry in the region near the paraflagellar body resemble that of a flavoprotein (Figure 2.16b). When *Euglena* goes from dark-adaptation to light-adaptation, the absorption spectrum maximal near 490 nm bleaches and the major absorption peak shifts to 440 nm (Figure 2.17). Such a shift in absorption spectrum is indicative of a photosensitive pigment whose behavior has similarities to the visual pigment rhodopsin (Figures 5.11 and 6.1). These studies suggest that in the photoreceptor system of *Euglena,* there are two pigments: one which acts as the primary photoreceptor molecule, while the other is a screening or accessory pigment in the process.

The flagellum, associated with the eyespot of protozoan flagellates (Figure 2.6), is also found as the connecting fiber between the inner and outer segments of vertebrate retinal rods (Figure 5.5), and such flagella or cilia are found associated with sensory cells throughout all phyla (Goss, 1970).

Analogies in photobehavior, photoreceptor structure, and pigments were drawn between the simple eyespot–flagellum photoreceptor system and the more complex photoreceptor system of arthropods and molluscs.

Visual Photoreceptors

In the arthropods (insects, arachnids, and crustacea) as well as in the molluscs, all kinds of optical imaging systems are found from pinhole camera-type eyes to ocelli, compound eyes, and refracting eyes (Figure 7.1).

The compound eye of arthropods is composed of numerous ommatidia. Each ommatidium consists of a corneal lens, a crystalline cone, and retinula cells that form a rhabdom (Figure 3.6). Each retinula cell has a differentiated photoreceptor structure, a rhabdomere, whose structure and function are similar to those of the vertebrate retinal rods.

Two anatomically and functionally distinct types of compound eyes, the apposition and the superposition, were described by Exner (1891). In the apposition eye the rhabdomeres that form the rhabdom lie directly beneath the crystalline cone (Figures 3.6a and 3.8). Superposition eyes

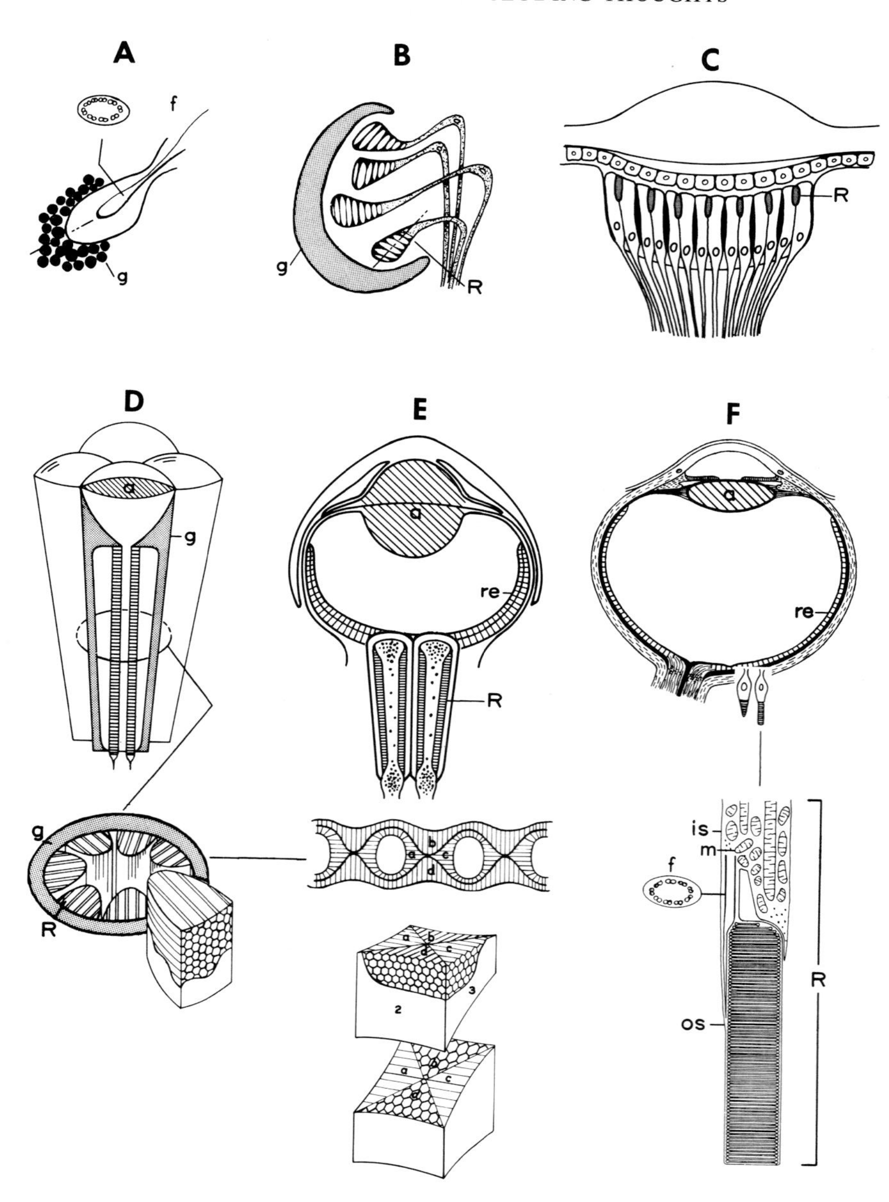

Figure 7.1. Schematic structure of various kinds of eyes and their photoreceptors. (A) Eyespot of *protozoa;* (B) photosensory cell of the flatworm ocellus; (C) insect ocellus; (D) compound eye of arthropods; (E) mollusc cephalopod eye; (F) vertebrate eye; a, lens, f, flagellum, g, pigment granules, R, photoreceptor (retinal rod or rhabdomere) os, outer segment, is, inner segment, m, mitochondria, re, retina.

are those in which the rhabdom lies some distance away from the crystalline cones (Figures 3.6b and 3.13).

An interesting structural observation in certain insects was that the crystalline cone and the rhabdom are connected by crystalline threads, as seen in the firefly, *Photuris* (Figure 3.13d). It was suggested that the crystalline threads function as waveguides (Horridge, 1969), and Døving and Miller (1969) showed that for the skipper butterfly, *Epargyreus clarus,* only the light contained in the crystalline thread is effective for stimulating the retinula cells. Varela and Wiitanen (1970) investigated the compound eye of the worker honeybee and showed that the closed rhabdom and the surrounding zone act together as a waveguide. These experimental findings raise doubts as to whether the superposition type compound eye is functional as envisaged by Exner (1891).

In the cephalopod mollusc the eye is a refracting type eye similar to the vertebrate eye (Figure 4.16), but its retina resembles that of the compound eye with retinula cells and their rhabdomeres that form rhabdoms (Figures 4.17–4.20).

Although much research remains before we can understand how these eyes function for vision, progress has been made in unraveling the rhabdom geometry and the molecular structure of the rhabdomere photoreceptors in a variety of invertebrate eyes.

In the insects and crustacea that have been described, there are two geometrical arrangements for the rhabdomeres that form the rhabdom. One is the "open-" type rhabdom as seen in *Drosophila* and *Copilia* (Figures 3.8c and 4.13d), in which the rhabdomeres project through a necklike part of their retinula cell into a cavity. The other is the "closed-" type or fused rhabdom as seen in the cockroach, firefly, and hornet, as well as the honeybee, grasshopper, locust, dragonfly, moth, and butterfly (Figures 3.10, 3.13–3.19), and in the crustacea (Figures 4.2 and 4.8). The closed-type rhabdom is also found in the cephalopod molluscs *Octopus,* cuttlefish, and squid (Figures 4.17 and 4.19).

The number of rhabdomeres that form the rhabdom vary; in the open rhabdom there are usually five to seven with an asymmetric rhabdomere lying in the same plane (Figures 3.8 and 4.13). In the closed rhabdom there are four to eight rhabdomeres forming a symmetrical arrangement, whereas its assymetric rhabdomere lies in another plane (Figures 3.14, 3.16, and 3.18). These two structural arrangements are schematically illustrated in Figures 3.9, 3.11, and 4.20.

Electrophysiological measurements indicate that there also may be two physiological types of eyes: a "slow-" type eye which is dependent upon the state of dark-adaptation, and a "fast-" type eye which is independent of the state of dark-adaptation (Autrum, 1950, 1958). It is of interest, then, that except for the adult dragonfly which has a closed rhabdom and fast response, most of the arthropods studied that have a closed rhabdom possess a slow electrical response. On the other hand, all the dipterous and hymenopterous insects that have an open rhabdom possess the fast-type electrical response. Generally the slow-type eye is more characteristic of nocturnal arthropods which would have a light gathering problem. As a consequence the closed rhabdom would greatly increase the surface area available for light capture by the visual pigment.

The closed rhabdom observed for the arthopods and molluscs is also found in the vertebrate amphibian retinal rod (Figure 5.4). Whether this is an evolutionary development or arose independently in the amphibians for greater light capture is not known.

The fine structure of all rhabdomeres is that of packed microtubules (microvilli) of about 500 Å in diameter whose double-walled membranes are about 100 Å in thickness (Figure 3.9). The microtubules also appear as lamellae, but this is dependent on the angle of cut through the rhabdomere. A reconstructed three-dimensional section of a rhabdomere shows how the microtubules are packed (Figure 7.2). In the rhabdom the microtubules of two adjacent rhabdomeres are oriented perpendicular to each other, whereas those of the opposite rhabdomeres the microtubules are parallel (Figures 3.9, 3.11, and 4.20).

In comparison, the vertebrate retinal rod and cone outer segments are electron dense lamellae, with a repeat unit of the order of 250 Å, and are separated by less dense interspaces (Figures 5.5 and 7.3a). The lamellae are double-membraned structures; each membrane is about 50 Å in thickness. Even though the vertebrate rods are described as lamellae and the rhabdomeres as microtubules, they are both membraneous processes of the retinal or retinula cells. These two types of membrane structures, microtubules for the invertebrates and lamellae for the vertebrates, depending on how the photoreceptor membranes are packed, will give either a tubular or lamellar structure. How these two kinds of structures can be developed are illustrated schematically in Figure 7.4.

This kind of molecular organization for the visual photoreceptors is by no means unique, for it has counterparts in the photosynthetic chloroplasts of plant cells (Figures 2.3b and 2.4c).

As we approach molecular dimensions for all photoreceptors, we see that there is a basic molecular structure. This structure has similarities to the lattice structure of crystals (Figure 7.5), which has led to the idea that the physics of photoreceptors may bear a close relation to the physics of the *solid state,* i.e., having similar properties of electronic energy transfer or electronic charge transfer. In electronic energy transfer, the

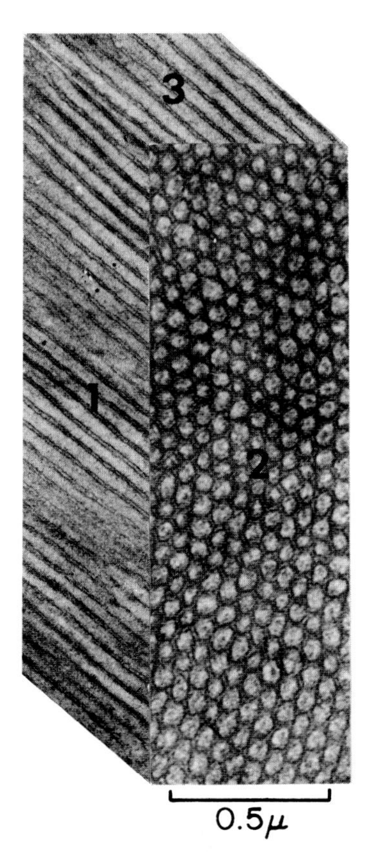

Figure 7.2. Reconstructed three-dimensional section of a rhabdomere from electron micrographs to show how their tubules are packed.

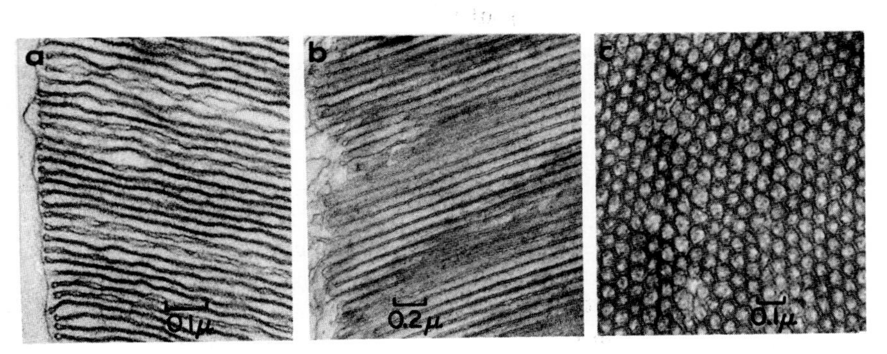

Figure 7.3. (a) Cattle retinal rod outer segment lamellae; (b) and (c) two different sections of a *Daphnia* rhabdomere to show lamellae and microtubules.

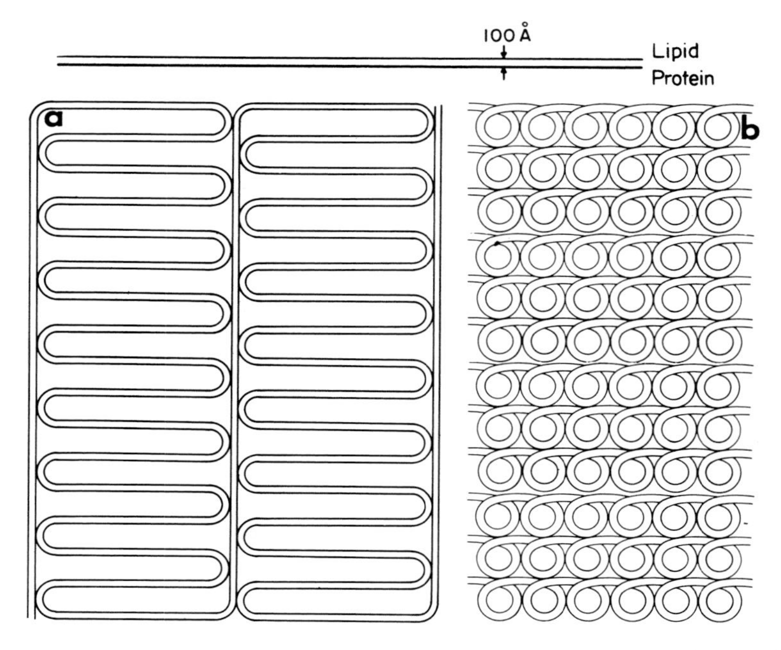

Figure 7.4. Schematic of how membranes can form (a) lamellae, or (b) microtubules.

light energy absorbed by one pigment molecule raises the molecule to an excited state. The energy may disappear from the first molecule and reappear on another molecule some distance away. This transfer of energy can continue until the energy reaches a molecule capable of triggering an electrical signal. In electronic charge transfer, the primary molecule that absorbs the light releases an electronic charge, which can move away from its original site if an electronic field is present, thus producing an electric current. Since the current would be caused by light absorption, the process is called *photoconductivity*.

In the formation of a charge transfer complex, the molecules involved must be brought close together, which can be done conveniently by building the substances into a crystal in which the lattice forces will hold the complex together. Here, then, is a clue to how photoreceptors function for energy transfer in the living state. If we interpolate this "crystalline" structure to the dimensions of a photoreceptor, we see that it provides a matrix for the orientation of the photosensitive pigment, brings the interacting molecules within molecular distances, and therefore serves as an efficiency mechanism for light capture and energy transfer.

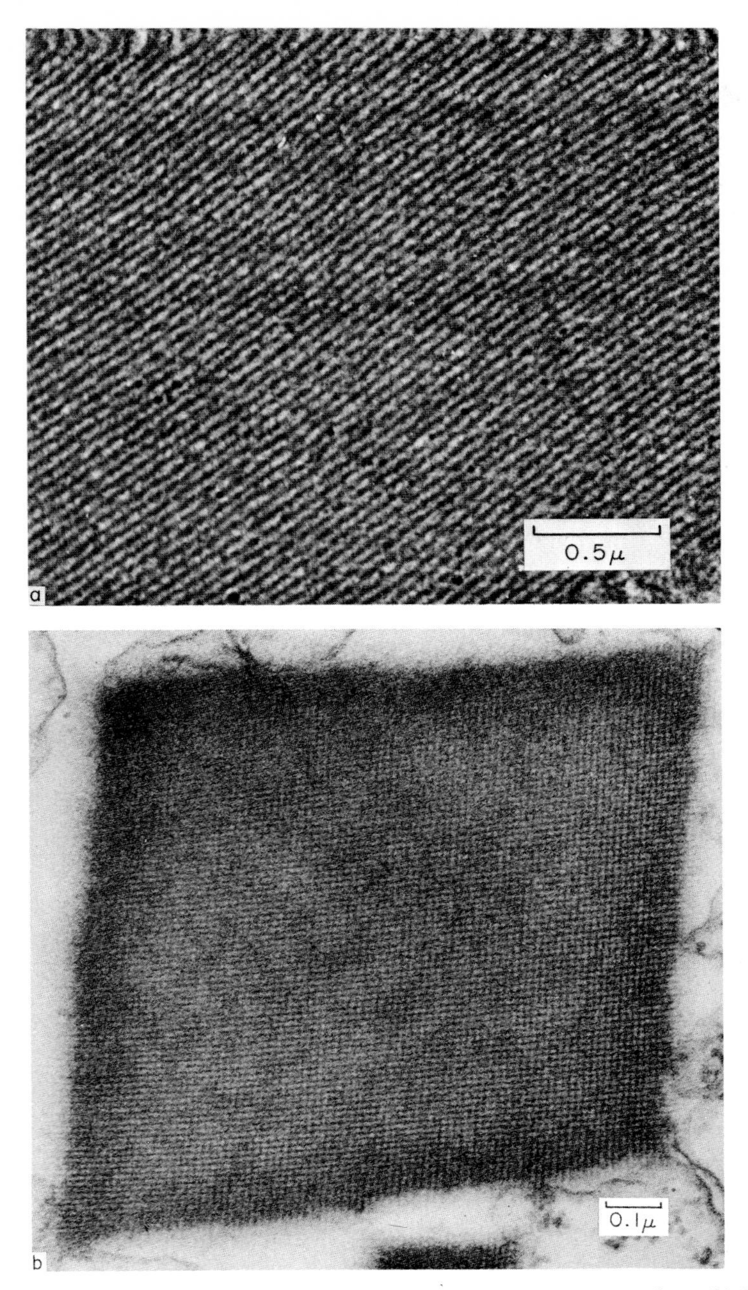

Figure 7.5. (a) The crystalline nature of a rhabdomere section, from *Daphnia*. (b) Crystal from photoreceptor area of *Phycomyces blakesleeanus*, a phototropic fungus (Wolken, 1969).

Photoreceptor Molecules

Not only do we find similarities in the molecular structure for all photoreceptors, but what may be equally important is the fact that all visual photoreceptors depend for their function on a single molecular group of pigments, the carotenoids and their derivatives (Figure 7.6).

All visual pigments (rhodopsins) thus far isolated from vertebrate and invertebrate eyes contain as the chromophore retinal$_1$ in the 11-*cis* geometric form. In the photochemistry of rhodopsin, retinal is released from the complexed form to the uncomplexed all-*trans*-retinal and opsin (Figures 5.12 and 5.13) Retinal is a linear molecule about 5 Å in diameter

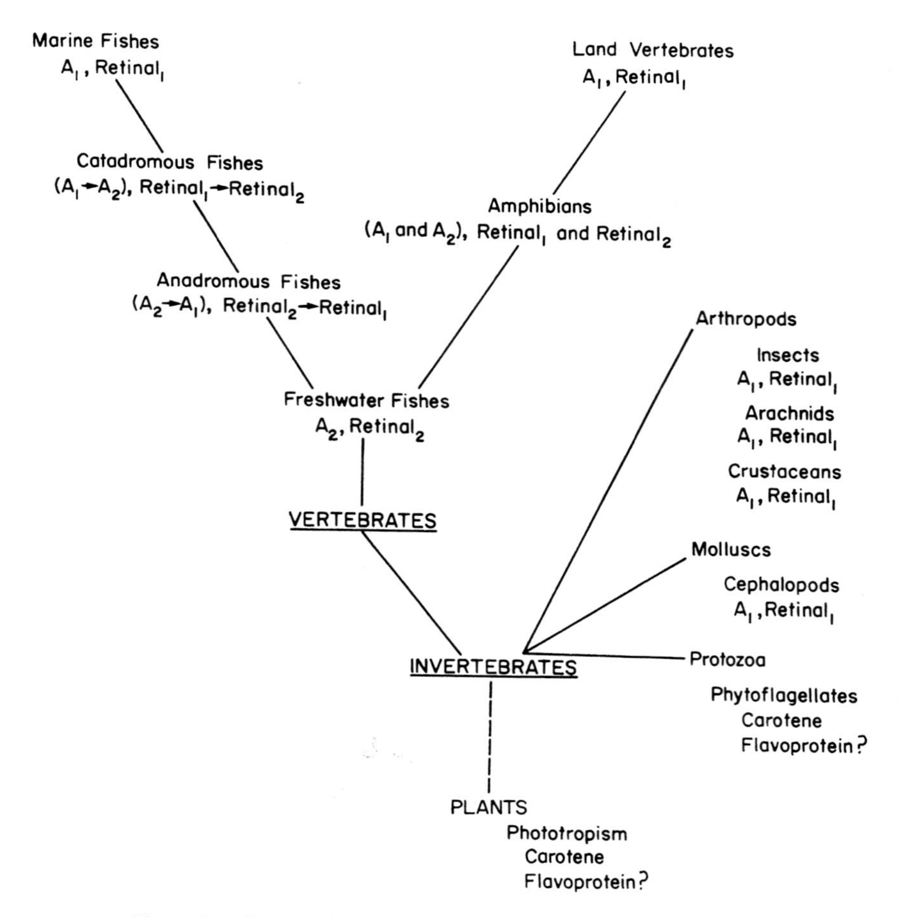

Figure 7.6. Carotenoid pigments associated with photoreceptors.

with a molecular weight of 280. When 11-*cis*-retinal is complexed with opsin to form rhodopsin it is about 40–50 Å in diameter (if a globular molecule) and its molecular weight is about 30,000–40,000 (Wolken, 1966). However, no real attempt has been made to estimate the size and molecular weight of invertebrate rhodopsins.

In the photochemistry of cephalopod molluscs and certain crustacea, rhodopsin does not bleach upon light absorption to retinal and opsin as do vertebrate rhodopsins (Figure 5.12). Instead, an equilibrium is established between rhodopsin and metarhodopsin (Figure 6.1a). The invertebrate metarhodopsins are pH indicators, being red (λ max near 500 nm) in neutral or slightly acid solution, and yellow (λ max near 380 nm) in alkaline solution (Figure 6.1a,b).

Insects have a typical rhodopsin, but in the housefly and the honeybee a visual pigment containing $retinal_1$, with an absorption maximum near 440 nm, is extractable with phosphate buffer (Figures 6.5 and 6.9). This photosensitive pigment differs from vertebrate and other invertebrate rhodopsins which require a solubilizing agent, e.g., digitonin, for extraction. Such a difference in solubility and absorption maximum from typical rhodopsins would indicate that there are differences in the protein, opsin, in these complexes.

The concentrations of rhodopsin that have been determined for arthropod and mollusc eyes are of the order of 10^7 molecules per rhabdomere, which is comparable to the concentration of rhodopsin molecules found for vertebrate retinal rods (Table 7.1). The number of rhodopsin molecules is directly related to the number of lamellae and their surface area in the retinal rod outer segment (Table 5.2), and similarly to the number of microtubules in the rhabdomere. We visualize, then, that rhodopsin is oriented on the surfaces of the retinal rod lamellar membranes, as a pigment lipoprotein (Figure 5.15). In an analogous manner, invertebrate rhodopsin would be oriented on the rhabdomere microtubule membranes.

Spectral Sensitivity, Pigments, and Color Vision

Behavioral studies indicate that many insects, crustacea, and molluscs can distinguish colors. Therefore, an important question to be answered is do these invertebrate eyes possess different spectrally absorbing pigments that would account for their ability to distinguish colors?

Color vision in vertebrates is known to be associated with the retinal cones. Using microspectrophotometry for obtaining spectra of individual cones, three different pigments were found (Marks *et al.,* 1964; Wald,

TABLE 7.1
CONCENTRATION OF VISUAL PIGMENT, RHODOPSIN, IN ARTHROPODS AND VERTEBRATES

		Average volume in cm^3	Concentration of rhodopsin molecules
Arthropods		rhabdomere	
Cockroach	*Periplaneta americana*	3.1×10^{-10}	5.9×10^{7}
Cockroach	*Blatta orientalis*	3.0×10^{-10}	4.3×10^{7}
Housefly	*Musca domestica*	6.8×10^{-11}	3.7×10^{7}
Honeybee	*Apis mellifera*	2.6×10^{-11}	2.6×10^{7}
Vertebrates		retinal rod	
Amphibian			
Frog	*Rana pipiens*	1.5×10^{-9}	3.0×10^{9}
Cattle		7.5×10^{-12}	1.0×10^{6}
Man		1.6×10^{-10}	1.0×10^{7}

1964). There are cones with blue sensitivity around 445 nm, cones with green sensitivity around 540 nm, and cones with red sensitivity around 560 nm. These absorption peaks coincide with the spectral sensitivity of the human eye (Wald, 1964) and are consistent with the psychophysical theories for color vision. However, using extraction procedures only one cone pigment has been isolated, either iodopsin at 562 nm when $retinal_1$ is present, or cyanopsin at 620 nm when $retinal_2$ is present (Figures 5.7 and 5.8).

Spectral sensitivity measurements for the invertebrates indicate that there are also several different absorbing pigments comparable to the rods and cones of vertebrate eyes. That is, they possess photoreceptors adapted for low and high levels of illumination and for color vision. For example, in crustacea, the lobster has visual pigments absorbing at 480 and 510 nm and the crayfish has visual pigments absorbing at 510 and 562 nm, that coincide with rod rhodopsin and cone iodopsin (Figure 5.7). In insects there are pigments with absorption peaks near 440 and 510 nm (Table 6.1). In addition to these absorption peaks there is an ultraviolet peak in the neighborhood of 340–390 nm.

There is still some question as to whether the behavioral action spectrum for insects, which shows responses in the red around 620 nm, reflects the absorption of a visual pigment (Marak and Wolken, 1965). According to Goldsmith (1965) and Strother (1966), such behavior is accounted for by the action of the screening pigments, for example, that of the reduced xanthommatin (Figure 6.19).

Burkhardt (1962) measured the spectral sensitivity of the blowfly,

Calliphora, using microelectrodes in single retinula cells. He found three different spectral sensitivity peaks with maxima at 470, 490, and 520 nm; each was coupled with an ultraviolet peak near 350 nm. In the worker bee, spectral sensitivity peaks were found at 340, 430, and 530 nm, and for the drone bee spectral sensitivity maxima were found at 340 and 447 nm (Autrum and von Zwehl, 1962; 1964).

The spectral sensitivities at 430 nm for the worker bee and 447 nm for the drone bee agree well with the extracted photosensitive honeybee pigment which has its maximum absorption near 440 nm (Figure 6.5). Langer and Thorell (1966) were able to obtain with the microspectrophotometer spectra of single rhabdomeres from the *Calliphora* eye mutant "chalky." The rhabdom of this eye has seven rhabdomeres of the open-type (see Figure 3.8). For six of these rhabdomeres the absorption maximum was 510 nm and the seventh or asymmetric rhabdomere the absorption maximum was at 470 nm. These spectral peaks come close to Burkhardt's maxima for spectral sensitivity, and presumably these absorption peaks are associated with two different visual pigments in different photoreceptors.

Even though spectral evidence would indicate several different absorbing pigments, there is the possibility that there is only one invertebrate rhodopsin visual pigment. The variations in absorption maxima could be the result of differences in the protein that forms the visual complex with retinal.

Behavioral action spectra and spectral sensitivity show a very strong ultraviolet response. The ultraviolet response is probably not due to retinal, but to another ultraviolet absorbing molecule. If the molecule which absorbs this ultraviolet light fluoresces in the visible, then its energy could be reabsorbed by the rhodopsin pigment (Chance, 1964). Such a possibility was discounted by De Voe *et al.* (1969) for the wolf spider, *Lycosa.* However, high ultraviolet sensitivities may indicate radiationless energy transfer (Hagins and Jennings, 1959); for example, from an ultraviolet absorbing accessory pigment molecule.

Accessory pigments such as flavines (Figure 1.11), pteridines, xanthommatins (Figure 6.19), and melanins are found associated with the screening pigment granules. Whether these molecules participate in the photoexcitation is not known at present.

Polarized Light Analysis

Sir John Lubbock (1882), an English banker, wondered how ants find their way to and from their nests while foraging for food. His observations led him to conclude that ants use the sun as a compass for orientation and navigation. Although this seemed to be a reasonable explanation, it was soon questioned—for how could insects find their way when the sun was hidden from view?

The question was not really answered until von Frisch's (1950) behavioral studies of the bee suggested that in addition to the direction of a point source, the sun, insects could utilize information by polarization from a patch of blue sky. In fact, there was an analyzer for polarized light in the bee's eye.

We have already discussed in Chapter IV the octagonal model of polarizing elements for the rhabdom that von Frisch (1953) proposed (Figure 4.21) to show that the different light transmission patterns obtained by looking at a patch of blue sky could account for the analysis of a beam of plane polarized light.

However, a number of other explanations for interpreting the mechanism of directional analysis of polarized light have been presented. One suggestion was a model which is dependent upon reflection–refraction at the air–corneal lens interface (Stephens *et al.,* 1953). Another was that the direction of the electric vector of a beam of plane polarized light could be perceived through simple intensity discriminations because the direction of vibration was resolved into intensity gradients by reflection from the background (Baylor and Smith, 1953). In fact, reflection patterns from the environment do resolve polarized light into patterns of graded intensity. Accordingly, the compound eye merely discriminates intensity and is not a direct analyzer (Kennedy and Baylor, 1961). In this context an interesting observation of von Frisch (1967) was that when a bee is dancing in the hive, the lower parts of the eyes would be used in perceiving reflection patterns, while the upper parts of the eye would be used if the stimulus was the direct perception of polarized light from the sky. von Frisch (1967) found that masking even a small area of the upper part of the eye disrupted dancing, while masking of the lower part of the eye had no effect.

With the information we now have on the invertebrate visual pigment, rhodopsin, the geometry of the rhabdoms and fine structure of the rhabdomeres, we can return to the question—where specifically is the analyzer for polarized light?

Dichroism in the Rhabdom

Recent ideas regarding the detection of polarized light by arthropods and molluscs suggest that the analyzer mechanism is based on the dichroism of the photosensitive pigment within the rhabdom (De Vries, 1956; Moody, 1964).

The basis for such a hypothesis goes back to the earlier studies of Schmidt (1934, 1935, 1938) on vertebrate retinal rod outer segments. He found that the absorption of polarized light is independent of the orientation of the electric vector for light traversing the rod parallel to its long axis, but is dichroic when irradiation is transverse. This dichroism was studied in more detail for marine retinae (Denton, 1959) and for frog retinal rods (Liebman, 1962). An interesting feature of these studies of dichroism in vertebrate rods is that when the dichroic ratio is plotted against wavelength it corresponds to the visual photosensitive pigment absorption spectrum. When the photosensitive pigment is bleached, the dichroism is lost. This would mean that at certain wavelengths, polarized light absorption would be good, while at other wavelengths it would not.

Burkhardt and Wendler (1960) recorded action potentials of single retinula cells in the compound eye of the blowfly, *Calliphora,* with intracellular electrodes. They found that rotating the plane of polarization resulted in a 50% difference in amplitude between maximum and minimum responses. This effect was obtained with blue and white light, but was not observed when red light was used as a stimulus. This finding that polarized blue and white light but not red light had an effect on the receptor potential may find its explanation in the dichroic absorption of oriented photopigments in the rhabdomere.

If, in fact, the rhabdom constitutes a dichroic analyzer, its properties would depend on the highly regular parallel arrangement of the microtubules in individual rhabdomeres, and within the membranes of the microtubules it would depend on the oriented rhodopsin molecules with their major axes parallel to the tubule direction and hence perpendicular to the normally incident illumination. Microspectrophotometry of the crayfish *(Orconectes virilis)* rhabdoms illuminated transversely does show that their photosensitive absorption exhibits a dichroic ratio of 2 *in situ*. The major absorption axis matches the axial direction of the closely parallel microtubules of the rhabdomere. Since these microtubules are regularly oriented transversely in about 24 layers, with the axes of the microtubules perpendicular to each other in alternate layers, transverse illumination of a properly oriented rhabdom displays alternate dichroic and isotropic bands. The explanation for the polarized light analyzer action in the rhabdom was given by Waterman *et al.* (1969). Their studies indicate that the absorbing dipoles of the rhodopsin molecules, as in the vertebrate retinal rods, lie parallel to the membrane surface (see Figure 5.15), but are otherwise randomly oriented. Thus they conclude that the rhabdom functional dichroism arises from the geometry of its fine structure.

Photoreceptor Evolution

Finally, a few thoughts are in order regarding the origins and development of photoreceptors. One hypothesis suggested here is that it was the ciliary and flagellar processes, developed by the cell for more rapid locomotion in an aqueous environment, to which photosensitive pigment globules became attached. Such a system provided a means for light searching as in the eyespot–flagellar mechanism of protozoan flagellates (Figures 2.6 and 2.7). Another suggestion is that photoreceptor development goes even further back than these specialized organelles to the origin of the cell membrane itself (Eakin, 1963, 1965, 1968).

As the invertebrate animals developed, the photoreceptors became restricted to certain regions of the epidermis which later invaginated forming simple eyes or ocelli. In examining these eyes (Figure 7.1), there seems to be no smooth phylogenetic development from eyespots and ocelli to compound and to refracting eyes, for the type of eye has been found to differ widely even among species of the same phylum.

Darwin (1859), wondering how the eye evolved from the photosensitive protozoa to an optical instrument as the vertebrate eye, wrote in *The Origin of Species:*

> When we reflect on these facts, here given much too briefly, with respect to the wide, diversified and graduated range of structure in the *eyes of lower animals;* and when we bear in mind how small the number of living forms must be in comparison with those which have become extinct, the difficulty ceases to be very great in believing that *natural selection* may have converted the simple apparatus of an optic nerve, coated with pigment and invested by transparent membrane, into an optical instrument . . .

However, the evolution of the eye continued to trouble him, and in writing to Asa Gray in 1860, Darwin wrote: "The eye to this day gives me a cold shudder . . ."*

It is no wonder that Darwin, with his keen sense for observation was troubled. It is only recently that we have been able to examine these photoreceptors with the tools of electron microscopy, x-ray diffraction, and spectroscopy. What we are now able to see is a new dimension and a new order in the visual photoreceptors. Two important observations stand out in these studies: one is that all photoreceptors use closely

*From F. Darwin (1892).

packed membranes, and secondly, that they contain a common photosensitive pigment molecule—a carotenoid or its degraded derivative (Figure 7.6). Wald (1957) expressed this as "natural selection on the molecular level."

Membranes hold a number of advantages which could be exploited for photoreception (Wolken, 1970). They provide an interface between the external and internal environments. Structurally, membranes are described as protein–lipid bilayers about 100 Å in thickness. Their structure may be associated with a unique property of lipids, that is, their ability to form monomolecular layers. This ability is attributable to the presence of a hydrophilic (water-soluble) group at one end of the molecule and a hydrophobic (fat-soluble) group at the other end. These membrane layers can serve to separate one part of a reaction from another. When folded into lamellae or tubes they minimize the receptor volume but greatly maximize the surface area available for all the photopigment molecules. Since the membranes are closely packed as in a crystal, they bring the receptor molecules within molecular distances for interaction.

To find the molecular basis for visual excitation will require more definitive experimentation and theoretical considerations. What has become apparent in summarizing these investigations is that membranes and photopigments form crystalline structures—and that such a structure is an efficiency mechanism for energy capture, amplification and regulation essential to the process of photoexcitation.

REFERENCES

Abrahamson, E. W. and Ostroy, S. E. (1967). The photochemical and macromolecular aspects of vision. *In* "Progress in Biophysics and Molecular Biology" (J. A. V. Butler and H. E. Huxley, eds.), Vol. XVII, pp. 179–215. Pergamon Press, Oxford.

Arnon, D. I. (1965). Ferredoxin and photosynthesis. *Science* **149,** 1460.

Astbury, W. T. and Saha, N. N. (1953). Structure of algal flagella. *Nature (London)* **171,** 280.

Autrum, H. (1950). Die Belechtungspotentiale und das Sehen der Insekten (Untersuchung an *Calliphora* und *Dixippus*). *Z. Vergl. Physiol.* **32,** 176.

Autrum, H. (1958). Electrophysiological analysis of the visual systems in insects. *Exp. Cell Res. Suppl.* **5,** 426.

Autrum, H. and Burkhardt, D. (1961). Spectral sensitivity of single visual cells. *Nature (London)* **190,** 639.

Autrum, H. and Langer, H. (1958). Photolabile Pterine im Auge von *Calliphora erythrocephala. Biol. Zentralbl.* **77,** 196.

Autrum, H. and Stumpf, H. (1950). Das Bienenauge als Analystor für polarisiertes Licht. *Z. Naturforsch.* **5b,** 116.

Autrum, H. and von Zwehl, V. (1962). Zur spektralen Empfindlichkeit einzelner Sehzellen der Drone *Apis mellifica. Z. Vergl. Physiol.* **46,** 8.

Autrum, H. and von Zwehl, V. (1964). Die spektrale Empfindlichkeit einzelner Sehzellen des Bienenauges. *Z. Vergl. Physiol.* **48,** 357.

Ball, S., Goodwin, T. W., and Morton, R. A. (1948). Studies in vitamin A. 5: The preparation of retinene$_1$–vitamin A aldehyde. *Biochem. J.* **42,** 516.

Ball, S., Collins, F. D., Dalvi, P. D., and Morton, R. A. (1949). Studies in vitamin A. 5: The preparation of retinene$_2$–vitamin A aldehyde. *Biochem. J.* **45,** 304.

Bamji, M. S. and Krinsky, N. I. (1965). Carotenoid de-expoxidations in algae expoxidations. Enzymatic conversion of antheraxanthin to zeaxanthin. *J. Biol. Chem.* **240,** 467.

Bartsch, R. G. and Kamen, M. D. (1960). Isolation and properties of two soluble heme proteins in extracts of photoanaerobe *Chromatium. J. Biol. Chem.* **235,** 825.

Batra, P. and Tollin, G. (1964). Phototaxis in *Euglena*. I. Isolation of the eyespot granules and identification of the eyespot pigments. *Biochim. Biophys. Acta* **79,** 371.

Baylor, E. R. and Smith, F. E. (1953). The orientation of *Cladocera* in polarized light. *Amer. Natur.* **87,** 97.

Beaven, G. H. and Holiday, E. R. (1952). Ultraviolet absorption spectra of proteins and amino acids. *Advan. Protein Chem.* **7,** 319.

Bernard, G. D. and Miller, W. H. (1968). Interference filters in the corneas of *Diptera. Invest. Ophthalmol.* **7,** 416.

Bertholf, L. M. (1931). The distribution of stimulative efficiency in the ultraviolet spectrum of the honeybee. *J. Agr. Res.* **43,** 703.

Blaurock, A. E. and Wilkins, M. H. F. (1969). Structure of frog photoreceptor membranes. *Nature (London)* **223,** 906.

Bliss, A. F. (1943). Derived photosensitive pigments from invertebrate eyes. *J. Gen. Physiol.* **26,** 361.

Bliss, A. F. (1948). The absorption spectra of visual purple of the squid and its bleaching products. *J. Biol. Chem.* **176,** 563.

Bowness, J. M. (1959). Preparation of rhodopsin using columns containing calcium triphosphate. *Biochim. Biophys. Acta* **31,** 305.

Bowness, J. M. and Wolken, J. J. (1959). A light sensitive yellow pigment from the housefly. *J. Gen. Physiol.* **42,** 779.

Brown, P. K. and Brown, P. A. (1958). Visual pigments of the octopus and cuttlefish. *Nature (London)* **182,** 1288.

Brown, P. K. and Wald, G. (1964). Visual pigments in single rods and cones of the human retina. *Science* **144,** 45.

Bünning, E. and Schneiderhöhn, G. (1956). Über das Aktionspecktrum der phototakischen Reaktionen von *Euglena. Arch. Mikrobiol.* **24,** 80.

Buriain, H. M. and Ziv, B. (1959). Electric response of the phakic and aphakic human eye to stimulation with near ultraviolet. *Arch. Ophthamol.* **61,** 347.

Burkhardt, D. (1962). Spectral sensitivity and other response characteristics of single visual cells in the arthropod eye. *In* "Biological Receptor Mechanisms" (J. W. L. Beament, ed.), pp. 86–109. Academic Press, New York.

Burkhardt, D. and Wendler, L. (1960). Ein direkter Beweis für die Fähigkeit einzelner Sehzellen des Insektenauges, die Schwingungsrichtung polarisierten Lichtes zu analysieren. *Z. Vergl. Physiol.* **43,** 687.

Burtt, E. T. and Catton, W. T. (1962). A diffraction theory of insect vision. I. An experimental investigation of visual acuity and image formation in the compound eyes of three species of insects. *Proc. Roy. Soc. Ser.* **B157,** 53.

Butenandt, A. (1952). The mode of action of hereditary factors. *Endeavour* **11,** 188.

Butenandt, A. and Neubert, G. (1955). Über Ommochrome V. Xanthommatins ein Augenfarbstoff der Schmeissfliege. *Hoppe-Seyler's Z. Physiol. Chem.* **301,** 109.

Butenandt, A., Schiedt, V., and Bickert, E. (1954). Über Ommochrome. **III.** Mitteilung Synthese des Xanthommatins. *Justis Liebigs Ann. Chem.* **588,** 106.

Butler, W. L., Siegelman, H. W., and Miller, C. O. (1964). Denaturation of phytochrome. *Biochemistry* **3,** 851.

Cajal, S. R. (1918). Observaciónes sobre la estructura de los ocelos y vias nerviosas ocelares de algunos insectos. *Trab. Lab. Invest. Biol. Univ. Madrid* **16,** 109.

Caspersson, T. O. (1950). "Cell Growth and Cell Function." W. W. Norton, New York.

Chance, B. (1964). Fluorescence emission of mitochondrial DPNH as a factor in the ultraviolet sensitivity of visual receptors. *Proc. Nat. Acad. Sci. U.S.A.* **51,** 359.

Clayton, R. K. (1953). Studies in the phototaxis of *Rhodospirillum rubrum.* I. Action spectrum, growth in green light, and Weber Law adherence. *Arch. Mikrobiol.* **19,** 107.

Clayton, R. K. (1965). "Molecular Physics in Photosynthesis." Ginn (Blaisdell), Boston, Massachusetts.

Cohen, A. I. (1961). The fine structure of the extra-foveal receptors of the Rhesus monkey. *Exp. Eye Res.* **1,** 128.

Cohen, A. I. (1963a). The fine structure of the visual receptors of the pigeon. *Exp. Eye Res.* **2,** 88.

Cohen, A. I. (1963b). Vertebrate retinal cells and their organization. *Biol. Rev. Cambridge Phil. Soc.* **38,** 427.

Cohen, A. I. (1964). Some observations on the fine structure of the retinal receptors of the American gray squirrel. *Invest. Ophthalmol.* **3,** 198.

Collins, F. D. (1954). Chemistry of vision. *Biol. Rev. Cambridge Phil. Soc.* **29,** 453.

Dartnall, H. J. A. (1957). "The Visual Pigments." Wiley, New York.

Dartnall, H. J. A. (1962). Photobiology of the visual process, Part II. *In* "The Eye" (Hugh Davson, ed.), Vol. II, pp. 323–522. Academic Press, New York.

Darwin, C. (1859). "The Origin of Species and the Descent of Man." pp. 133–135. Modern Library, New York, 1936.

Darwin, F., ed. (1892). "The Autobiography of Charles Darwin and Selected Letters." D. Appleton, New York; republished 1958, Dover, New York, p. 220.

Daumer, K. (1956). Reizmetrische Untersuchung des Farbensehens der Biene. *Z. Vergl. Physiol.* **38,** 413.

Delbrück, M. and Reichart, W. (1956). System analysis for the light growth reactions in *Phycomyces. In* "Cellular Mechanisms in Differentiation and Growth" (D. Rudnick, ed.), pp. 3–44. Princeton Univ. Press, Princeton, New Jersey.

Denton, E. J. (1959). The contributions of the oriented photosensitive and other molecules to the absorption of whole retina. *Proc. Roy. Soc. Ser.* **B150,** 78.

De Robertis, E. (1956). Electron microscope observations on the submicroscopic organization of retinal rods. *J. Biophys. Biochem. Cytol.* **2,** 319.

De Robertis, E. and Lasansky, A. (1961). Ultrastructure and chemical organization of photoreceptors. *In* "The Structure of the Eye" (G. K. Smelser, ed.), p. 29. Academic Press, New York.

De Voe, R. D., Small, R. J. W., and Zvagulis, J. E. (1969). Spectral sensitivities of wolf spider eyes. *J. Gen. Physiol.* **54,** 1.

De Vries, H. (1956). Physical aspects of the sense organs. *In* "Progress in Biophysics" (J. A. V. Butler, ed.), Vol. 6, p. 246. Pergamon Press, Oxford.

Diehn, B. and Tollin, G. (1966). Phototaxis in *Euglena.* II. Physical factors determining the rate of phototactic response. *Photochem. Photobiol.* **5,** 523.

Dietrich, W. (1909). Die Facettenaugen der Dipteran. *Z. Wiss. Zool.* **92,** 465.

Dodt, E. and Walther, J. B. (1958). Fluorescence of the crystalline lens and electroretinographic sensitivity determinations. *Nature (London)* **181,** 286.

Døving, K. B. and Miller, W. H. (1969). Function of insect compound eyes containing crystalline tracts. *J. Gen. Physiol.* **54,** 250.

Dowling, J. E. (1965). Foveal receptors of the monkey retina; fine structure. *Science* **147,** 57.

Eakin, R. M. (1963). Lines of evolution of photoreceptors. *In* "General Physiology of Cell Specialization" (D. Mazia and A. Tyler, eds.), pp. 393–425. McGraw-Hill, New York.

Eakin, R. M. (1965). Evolution of photoreceptors. Cold Spring Harbor Symp. Quant. Biol. **30,** 363.

Eakin, R. M. (1968). Evolution of photoreceptors. *In* "Evolutionary Biology" (T. Dobzhansky, M. K. Hecht, and W. C. Steere, eds.), Vol. II, pp. 194–242. Appleton-Century-Crofts, New York.

Eguchi, E. and Waterman, T. H. (1966). Fine structure patterns in crustacean rhabdoms. *In* "Functional Organization of the Compound Eye" (C. G. Bernhard, ed.), pp. 105–124. Pergamon Press, Oxford.

Eichenbaum, D. M. and Goldsmith, T. H. (1968). Properties of intact photoreceptor cells lacking synapses. *J. Exp. Zool.* **169,** 15.

Englemann, T. W. (1882). Ueber Licht- und Farbenperception niederster Organismen. *Pflügers Arch. Gesampte Physiol. Menschen Tiere* **29,** 387.

Exner, S. (1891). "Die Physiologie der Facettierten Augen von Krebsen und Insekten." Leipzig, Germany and Franz Deuticke, Vienna, Austria.

Fauré-Frémiet, E. (1958). The origin of the metazoa and the stigma of the phytoflagellates. *Quart. J. Microsc. Sci.* **99,** 123.

Fauré-Frémiet, E. and Rouiller, C. (1957). Le flagelle interne d'une Chrysomondale: *Chromulina psammobia*. *C. R. Soc. Biol.* **244,** 2655.

Fawcett, D. W. and Porter, K. R. (1952). A study of the fine structure of ciliated epithelial cells with the electron microscope. *Anat. Rec.* **113,** 539.

Fawcett, D. W. and Porter, K. R. (1954). A study of the fine structure of ciliated epithelia. *J. Morpho.* **94,** 221.

Fernández-Morán, H. (1956). Fine structure of the insect retinula as revealed by electron microscopy. *Nature (London)* **177,** 742.

Fernández-Morán, H. (1958). Fine structure of the light receptors in the eyes of insects. *Exp. Cell Res. Suppl.* **5,** 586.

Fernández-Morán, H. (1961). The fine structure of vertebrate and invertebrate photoreceptors as revealed by low-temperature electron microscopy. *In* "The Structure of the Eye" (G. K. Smelser, ed.), pp. 521–556. Academic Press, New York.

Fingerman, M. (1952). The role of the eye pigments of *Drosophila melanogaster* in photic orientation. *J. Exp. Zool.* **120,** 131.

Fingerman, M. and Brown, F. A. (1952). A "Purkinje shift" in insect vision. *Science* **116,** 171.

Fingerman, M. and Brown, F. A. (1953). Color discrimination and physiological duplicity of *Drosophila* vision. *Physiol. Zool.* **26,** 59.

Forrest, H. S. and Mitchell, H. K. (1954a). The pteridines of *Drosophila melanogaster. Chem. Biol. Pteridines, Ciba Found. Symp., 1954,* p. 143.

Forrest, H. S. and Mitchell, H. K. (1954b). Pteridines from *Drosophila.* I. Isolation of a yellow pigment. *J. Amer. Chem. Soc.* **76,** 5656.

Fox, D. L. (1953). "Animal Biochromes and Structural Colors." Cambridge Univ. Press, London and New York.

Fraenkel, G. S. and Gunn, D. L. (1961). "The Orientation of Animals: Kinesis, Taxis, and Compass Reactions." Dover, New York.

Gascoigne, S. C. B. (1968). The optics of large telescopes. *Quart. J. Roy. Astron. Soc.* **9,** 98.

Gerschler, M. W. (1911). Monographie du *Leptodora kindtii* (Focke), *Arch. Hydrobiol. Planktonk.* **6,** 415.

Gerschler, M. W. (1912). Monographie du *Leptodora kindtii* (Focke), *Arch. Hydrobiol. Planktonk.* **7,** 63.

Gössel, I. (1957). Über das Aktionspektrum der Phototaxis chlorophyllfreier Euglenen und über die Absorption des Augenflecks. *Arch. Mikrobiol.* **27,** 288.

Goldsmith, T. H. (1958a). The visual system of the honeybee. *Proc. Nat. Acad. Sci. U.S.A.* **44,** 123.

Goldsmith, T. H. (1958b). On the visual system of the bee *(Apis mellifera). Ann. N. Y. Acad. Sci.* **74,** 223.

Goldsmith, T. H. (1960). The nature of the retinal action potential and the spectral sensitivities of ultraviolet and green receptor systems of the compound eye of the worker honeybee. *J. Gen. Physiol.* **43,** 775.

Goldsmith, T. H. (1962). Fine structure of the retinulae in the compound eye of the honeybee. *J. Cell Biol.* **14,** 489.

Goldsmith, T. H. (1964). The visual system of insects. *In* "The Physiology of Insects" (M. Rockstein, ed.), Vol. I, pp. 394–462. Academic Press, New York.

Goldsmith, T. H. (1965). Do flies have a red receptor? *J. Gen. Physiol.* **49,** 265.

Goldsmith, T. H. and Philpott, D. E. (1957). The microstructure of the compound eyes of insects. *J. Biophys. Biochem. Cytol.* **3,** 429.

Goldsmith, T. H. and Warner, L. T. (1964). Vitamin A in the vision of insects. *J. Gen. Physiol.* **47,** 433.

Goldsmith, T. H., Dizon, A. E., and Fernández, H. R. (1968). Photoreceptor organelles of the prawn *Palaemonetes*. *Science* **161,** 468.

Goodwin, T. W. (1952). "Comparative Biochemistry of Carotenoids." Chapman and Hall, London.

Goodwin, T. W., ed. (1965). "Chemistry and Biochemistry of Plant Pigments." Academic Press, New York.

Goodwin, T. W. and Gross, J. A. (1958). Carotenoid distribution in bleached substrains of *Euglena gracilis*. *J. Protozool.* **5,** 292.

Goodwin, T. W. and Jamikorn, M. (1954). Studies in carotenogenesis. Some observations on carotenoid synthesis in two varieties of *Euglena gracilis*. *J. Protozool.* **1,** 216.

Goss, R. J. (1970). Turnover in cells and tissues. *In* "Advances in Cell Biology" (D. M. Prescott, L. Goldstein, and E. McConkey, eds.) Vol. I, pp. 233–296. Appleton-Century-Crofts, New York.

Granick, S. (1948). Magnesium protoporphyrin as a precursor of chlorophyll in *Chlorella*. *J. Biol. Chem.* **175,** 333.

Granick, S. (1950). Magnesium vinyl pheoporphyrin a_5, another intermediate in the biological synthesis of chlorophyll. *J. Biol. Chem.* **183,** 713.

Granick, S. (1958). Porphyrin biosynthesis in erythrocytes. I. Formation of ζ-aminolevulinic acid in erythrocytes. *J. Biol. Chem.* **232,** 1101.

Gregory, R. L. (1966). "Eye and Brain." McGraw-Hill, New York.

Gregory, R. L. (1967). Origin of eyes and brains. *Nature (London)* **213,** 5074.

Gregory, R. L., Morey, N., and Ross, H. E. (1964). The curious eye of *Copilia*. *Nature (London)* **201,** 1166.

Grenacher, H. (1879). "Untersuchungen über das Sehorgan der Arthropoden, insbesondere der Spinnen, Insekten und Crustaceen," p. 195. Vandenhoeck and Ruprecht, Göttingen, Germany.

Grenacher, H. (1886). Abhandlungen zur vergleichenden Anatomie des Auges. I. Die Retina der Cephalopoden. *Abh. Naturforsch. Ges. Halle* **16,** 207.

Grossbach, U. (1957). Zur papierchromatographischen Untersuchung von Lepidopterenaugen. *Z. Naturforsch.* **12b,** 462.

Guzzo, A. V. and Pool, G. L. (1969). Fluorescence spectra of the intermediates of rhodopsin bleaching. *Photochem. Photobiol.* **9,** 565.

Hagins, W. A. and Jennings, W. H. (1959). Radiationless migration of electronic excitation in retinal rods. *Discuss. Faraday Soc.* **27,** 180.

Halldal, P. (1964). Phototaxis in protozoa. *In* "Biochemistry and Physiology of Protozoa" (S. H. Hutner, ed.), Vol. III, pp. 277–296. Academic Press, New York.

Hanaoka, T. and Fujimoto, K. (1957). Absorption spectrum of a single cone in the carp retina. *Jap. J. Physiol.* **7,** 276.

Hartmann, K. M. (1966). A general hypothesis to interpret "high energy phenomena" of photomorphogenesis on the basis of phytochrome. *Photochem. Photobiol.* **5,** 349.

Hawkins, E. G. E. and Hunter, R. F. (1944). Vitamin A aldehyde. *J. Chem. Soc.* p. 411

Hays, D. and Goldsmith, T. H. (1969). Microspectrophotometry of the visual pigment of the spider crab. *Libinia emarginata*. *Z. Vergl. Physiol.* **65,** 218.

Heller, J. (1968). Structure of visual pigments. I. Purification, molecular weight, and composition of bovine visual pigment$_{500}$. *Biochemistry* **7,** 2906.

Heller, J. (1969). Comparative study of a membrane protein characterization of bovine, rat and frog visual pigment$_{500}$. *Biochemistry* **8,** 675.

Hendricks, S. B. (1968). How light interacts with living matter. *Sci. Amer.* **219,** No. 3, 174.

Hering, E. (1885). "Ueber Individuelle Verschiedenheiten des Farbensinnes." Lotos, Prague.

Hering, E. (1965). "Outlines of a Theory of the Light Sense" (Translated by L. M. Hurvich and D. Jameson). Harvard University Press, Cambridge, Massachusetts.

Hertz, M. (1939). New experiments on color vision in bees. *J. Exp. Biol.* **16,** 1.

Hess, W. N. (1943). Visual organs of invertebrate animals. *Sci. Mon.* **57,** 489.

Holmes, S. J. (1903). Phototaxis in *Volvox. Biol. Bull.* **4,** 319.

Holwill, M. E. J. (1966). Physical aspects of flagellar movement. *Physiol. Rev.* **46,** 696.

Hotta, Y. and Benzer, S. (1969). Abnormal electroretinograms in visual mutants of *Drosophila. Nature (London)* **222,** 354.

Horridge, G. A. (1966). The retina of the locust. *In* "The Functional Organization of the Compound Eye." (C. G. Bernhard, ed.), p. 513. Pergamon Press, Oxford.

Horridge, G. A. (1968). Pigment movement and the crystalline threads of the firefly eye. *Nature (London)* **218,** 778.

Horridge, G. A. (1969). The eye of the firefly *Photuris. Proc. Roy. Soc. Ser.* **B171,** 445.

Hubbard, R. (1954). The molecular weight of rhodopsin and the nature of the rhodopsin–digitonin complex. *J. Gen. Physiol.* **37,** 381.

Hubbard, R. and Kropf, A. (1959). Molecular aspects of visual excitation. *Ann. N. Y. Acad. Sci.* **81,** 442.

Hubbard, R. and St. George, R. C. C. (1958). The rhodopsin system of the squid. *J. Gen. Physiol.* **41,** 501.

Hubbard, R. and Wald, G. (1960). Visual pigment of the horseshoe crab *Limulus polyphemus. Nature (London)* **186,** 212.

Hunter, R. F. and Williams, N. E. (1945). Chemical conversion of β-carotene into vitamin A. *J. Chem. Soc.* p. 554.

Hutner, S. H. (1955). Comparative biochemistry of flagellates. *In* "Biochemistry and Physiology of Protozoa" (S. H. Hutner and A. Lwoff, eds.), Vol. II, pp. 1–40. Academic Press, New York.

Hutner, S. H. and Provasoli, L. (1951). The phytoflagellates. *In* "Biochemistry and Physiology of Protozoa." (A. Lwoff, ed.), Vol. I, pp. 27–128. Academic Press, New York.

Johnson, L. P. and Jahn, T. L. (1942). Cause of the green-red color change in *Euglena rubra. Physiol. Zool.* **15,** 89.

Jörschke, H. (1914). Die Facettenaugen der Orthopteren und Termiten. *Z. Wiss. Zool.* **111,** 153.

Kästner, A. (1950). Reaktion der Hupfspinnen (Salticidae) auf unbewegte farblose und farbige Gesichtsriege. *Zool. Beitr.* **1**[N.S.], 12.

Kamen, M. D. (1956). Hematin compounds in metabolism of photosynthetic tissues. *In* "Enzymes: Units of Biological Structure and Function" (O. H. Gaebler, ed.), p. 483. Academic Press, New York.

Kamen, M. D. (1960). Hematin compounds in photosynthesis. *In* "Comparative Biochemistry of Photoreactive Systems" (M. B. Allen, ed.), pp. 323–327. Academic Press, New York.

Kampa, E. M. (1955). Euphausiopsin. A new photosensitive pigment from the eyes of the euphausiid crustaceans. *Nature (London)* **175,** 996.

Karrer, P. and Jucker, E. (1950). "Carotenoids." Elsevier, New York.

Keeble, F. (1910). "Plant–Animal: A Study in Symbiosis." Cambridge Univ. Press, London and New York.

Kennedy, D. (1963). Physiology of photoreceptor neurons in the abdominal nerve cord of the crayfish. *J. Gen. Physiol.* **46,** 551.

Kennedy, D. and Baylor, E. R. (1961). Analysis of polarized light by the bee's eye. *Nature (London)* **191,** 34.

Kennedy, D. and Bruno, M. S. (1961). The spectral sensitivity of crayfish and lobster vision. *J. Gen. Physiol.* **44,** 1089.

Kikkawa, H. (1941). Mechanism of pigment formation in *Bombyx* and *Drosophila*. *Genetics* **26,** 587.

Kikkawa, H., Ogita, Z., and Fujito, S. (1955). Nature of pigments derived from tyrosine and tryptophan in animals. *Science* **121,** 43.

Kimbel, R. L. Jr., Poincelot, R. P., and Abrahamson, E. W. (1970). Chromophore transfer from lipid to protein in bovine rhodopsin. *Biochemistry* **9,** 1817.

Krinsky, N. I. (1964). Carotenoid de-expoxidations in algae. I. Photochemical transformation of antheraxanthin to zeaxanthin. *Biochim. Biophys. Acta* **88,** 487.

Krinsky, N. I. and Goldsmith, T. H. (1960). Carotenoids of *Euglena gracilis* (Z strain). *Fed. Proc. Fed. Amer. Soc. Exp. Biol.* **19,** 329.

Krukenberg, C. F. W. (1882). Ueber die Stäbchenfarbe der Cephalopoden. *Untersuch. Physiol. Inst. Heidelberg* **2,** 58.

Kuiper, J. W. (1962). The optics of the compound eye. *In* "Biological Receptor Mechanisms" (J. W. L. Beament, ed.), p. 58. Academic Press, New York.

Kuiper, J. W. (1966). On the image formation in a single ommatidium of the compound eye in diptera. *In* "Functional Organization of the Compound Eye" (C. G. Bernhard, ed.), pp. 35–50. Pergamon Press, Oxford.

Kühn, A. (1927). Über den Farbensinn der Bienen. *Z. Vergl. Physiol.* **5,** 762.

Kuwabara, M. (1957). Bildung des bedingten Reflexes von Paulous Typus bei der Honigbiene *Apis mellifera*. *J. Fac. Sci. Hokkaido Univ. Ser. 6, Zool.* **13,** 458.

Land, E. J. and Swallow, A. J. (1969). One-electron reactions in biochemical systems as studied by pulse radialysis. II. Riboflavin. *J. Biochem.* **8,** 2117.

Lane, F. W. (1960). "Kingdom of the Octopus." Pyramid, New York.

Langer, H. (1967). Über die Pigmentgranula im Facettenaugen von *Callifora erythrocephala*. *Z. Vergl. Physiol.* **55,** 354.

Langer, H. and Thorell, B. (1966). Microspectrophotometry of single rhabdomeres in the insect eye. *Exp. Cell Res.* **44,** 673.

Lasansky, A. (1967). Cell junctions in ommatidium of Limulus. *J. Cell. Biol.* **33,** 365.

Lee, J. W. (1954a). The effect of pH on forward swimming in *Euglena* and *Chilomonas*. *Physiol. Zool.* **27,** 272.

Lee, J. W. (1954b). The effect of temperature on forward swimming in *Euglena* and *Chilomonas*. *Physiol. Zool.* **27,** 275.

Lewin, R. A. (1955). Flagella: variations and enigmas. *New Biol.* **19,** 27.

Lewin, R. A. (1962). "Physiology and Biochemistry of Algae." Academic Press, New York.

Liebman, P. A. (1962). *In situ* microspectrophotometric studies on the pigments of single retinal rods. *Biophys. J.* **2,** 161.

Liebman, P. A. (1969). Microspectrophotometry of retinal cells. *Ann. N. Y. Acad. Sci.* **157,** 250.

Liebman, P. A. and Entine, G. (1964). Sensitive low-light level of microspectrophotometric detection of photosensitive pigment of retinal cones. *J. Opt. Soc. Amer.* **54,** 1451.

Linzen, B. (1959). Über die Verbreitung der Ommochrome, der dunkeln Augenpigments der Arthropoden. *Zool. Anz. Suppl.* **22,** 154.

Loeb, J. (1918). "Forced Movements, Tropism and Animal Conduct." Lippincott, Philadelphia, Pennsylvania.

Lubbock, J. (1882). "Ants, Bees and Wasps." D. Appleton and Co., New York.

MacNichol, E. F. (1964). Three pigment color vision. *Sci. Amer.* **211,** No. 6, 48.

Manten, A. (1948a). Phototaxis in the purple bacterium *Rhodospirillum rubrum* and the relation between phototaxis and photosynthesis. *Antonie van Leeuwenhoek Microbiol. Serol.* **14,** 65.

Manten, A. (1948b). Phototaxis, phototropism and photosynthesis in purple bacteria and blue-green algae. Thesis, Utrecht, Holland, Drukkerij Fa. Schotanus and Jens.

Manton, I. (1952). The fine structure of plant cilia. *Symp. Soc. Exp. Biol.* **6,** 306.

Marak, G. E., Jr. and Wolken, J. J. (1965). The action spectrum for the fire ant. *(Solenopsis saevissima). Nature (London)* **205,** 1328.

Marks, W. B. (1963). Difference spectra of the visual pigments in single goldfish cones. Thesis, Johns Hopkins University, Baltimore, Maryland.

Marks, W. B., Dobelle, W. H., and MacNichol, J. R. (1964). Visual pigments of single primate cones. *Science* **143,** 1181.

Mast, S. O. (1911). "Light and the Behavior of Organisms." Wiley, New York.

Mathews, R. G., Hubbard, R., Brown, P. K., and Wald, G. (1963). Tautomeric forms of metarhodopsin. *J. Gen. Physiol.* **47,** 215.

Maxwell, J. C. (1861). On the theory of compound colours and the relations of the colours of the spectrum. *Phil. Trans. Roy. Soc. London* **150,** 57.

Maxwell, J. C. (1890). On the theory of compound colours and the relations of the colours of the spectrum. *Sci. Papers,* Cambridge, **I,** 410.

Miller, W. H. and Bernard, G. D. (1968). Butterfly glow. *J. Ultrastruct. Res.* **24,** 286.

Miller, W. H., Bernard, G. D., and Allen, J. L. (1968). The optics of insect compound eyes. *Science* **162,** 760.

Millot, N. (1968). The dermal light sense. *In* "Invertebrate Receptors" (J. D. Carthy and G. E. Newell, eds.), pp. 1–36. Academic Press, New York.

Missotten, L. (1964). L'ultrastructure des tissues oculaires. *Bull. Soc. Belge Opthalmol.* **135,** 1.

Moody, M. F. (1964). Photoreceptor organelles in animals. *Biol. Rev. Cambridge Phil. Soc.* **39,** 43.

Morton, R. A. (1944). Chemical aspects of the visual process. *Nature (London)* **153,** 69.

Morton, R. A. and Goodwin, T. W. (1944). Preparation of retinene *in vitro. Nature (London)* **153,** 405.

Morton, R. A. and Pitt, G. A. J. (1955). Studies on rhodopsin. 9. pH and the hydrolysis of indicator yellow. *Biochem. J.* **59,** 128.

Mote, M. I. and T. H. Goldsmith. (1970). Spectral sensitivities of color receptors in the compound eye of the cockroach *Periplaneta. J. Exp. Zool.* **173,** 137.

Naka, K. I. (1960). Recording of retinal action potentials from single cells in the insect compound eye. *J. Gen. Physiol.* **44,** 571.

Nolte, D. J. (1950). The eye-pigmentary system of *Drosophila:* The pigment cells. *J. Genet.* **50,** 79.

Nowikoff, M. (1932). Ueber den Bau der Komplexaugen von *Periplaneta (Stylopyga) orientalis. Jena Z. Naturwiss.* **67,** 58.

Packard, A. and Sanders, G. (1969). What the octopus shows the world. *Endeavour* **28,** 92.

Patten, W. (1887). Eyes of molluscs and arthropods. *J. Morphol.* **1,** 67.

Peckman, G. W. and Peckman, E. G. (1887). Some observations on the mental powers of spiders. *J. Morphol.* **1,** 383.

Perralet, A. and Baumann, F. (1969). Evidence for extracellular space in the rhabdom of the honeybee drone eye. *J. Cell Biol.* **3,** 825.

Pitelka, D. R. (1963). "Electron-Microscope Structure of Protozoa." Pergamon Press, Oxford.

Pitelka, D. R. (1969). Fibriller systems in protozoa. *In* "Research in Protozoology II" (T. T. Chen, ed.), pp. 279–388. Pergamon Press, Oxford.

Pitelka, D. R. and Schooley, C. N. (1955). "Comparative Morphology of Some Protistan Flagella." Univ. of California Press, Berkeley, California.

Poincelot, R. P. and Abrahamson, E. W. (1970). Phospholipid composition and extractability of bovine rod outer segments and rhodopsin micelles. *Biochemistry* **9,** 1820.

Post, C. T. and Goldsmith, T. H. (1965). Pigment migration and light adaptation in the eye of the moth, *Galleria mellonella. Biol. Bull.* **128,** 473.

Pringsheim, E. G. (1963). "Farblose Algen. Ein Beitrag zur Evolutionsforschung." Gustav Fischer, Stuttgart.

Ramsey, J. A. (1952). "A Physiological Approach to Lower Animals." Cambridge Univ. Press, London and New York.

Röhlich, P. and Törö, I. (1965). Fine structure of the compound eye of Daphnia in normal, dark- and strongly light-adapted state. *In* "The Eye Structure II. Symposium." (J. W. Rohen, ed.), pp. 175–186. F. K. Schattauer-Verlag, Stuttgart.

Röhlich, P. and Török, L. J. (1961). Electronenmikroscopische Untersuchungen des Auges von Planarien. *Z. Zellforsch. Mikrosk. Anat.* **54,** 362.

Rogers, G. L. (1962). A diffraction theory of insect vision. II. Theory and experiments with a simple model eye. *Proc. Roy. Soc. Ser.* **B157,** 83.

Rosenbaum, J. L. and Child, F. M. (1967). Flagellar regeneration in protozoan flagellates. *J. Cell Biol.* **34,** 345.

Rosenberg, B. (1958). Photoconductivity and the visual receptor processes. *J. Opt. Soc. Amer.* **48,** 581.

Runge, W. J. (1966). A recording microfluorospectrophotometer. *Science* **151,** 499.

Salah, M. K. and Morton, R. A. (1948). Crystalline retinene$_2$. *Biochem. J.* **43,** 6.

St. George, R. C. C. and Wald, G. (1949). The photosensitive pigment of the squid retina. *Biol. Bull.* **97,** 248.

Scharrer, E. (1964). A specialized trophospongium in large neurons of *Leptodora* (Crustacea). *Z. Zellforsch. Mikrosk. Ant.* **61,** 803.

Schmidt, W. J. (1934). Dichroismus des Aussengliedes der Stäbchenzellen der Froschnetzhaut, verursacht durch den Sehpurpur. *Naturwissenschaften* **22,** 206.

Schmidt, W. J. (1935). Doppelbrechung, Dichroismus und Feinbau des Aussengliedes der Sehzellen vom Frosch. *Z. Zellforsch. Mikrosk. Anat.* **22,** 485.

Schmidt, W. J. (1937). Die Doppelbrechung von Karyoplasma, Zytoplasma, und Metaplasma, *Protoplasma Monogr.* **11.**

Schmidt, W. J. (1938). Polarisationsoptische Analyse eines Eiweiss-Lipoid-Systems, erläutert am Aussenglied der Sehzellen. *Kolloid Z.* **85,** 137.

Schultze, M. J. (1866). Zur Anatomie und Physiologie der Retina. *Arch. Mikrosk. Anat.* **2,** 175.

Shaw, S. R. (1969). Optics of arthropod compound eye. *Science* **165,** 88.

Shichi, H. (1970). Spectrum and purity of bovine rhodopsin. *Biochemistry* **9,** 1973.

Siegelman, H. W. and Firer, E. M. (1964). Purification of phytochrome from oat seedlings. *Biochemistry* **3,** 418.

Sjöstrand, F. S. (1949). An electron microscope study of the retinal rods of the guinea pig eye. *J. Cell. Comp. Physiol.* **33,** 383.

Sjöstrand, F. S. (1953a). The ultrastructure of the outer segments of rods and cones of the eye as revealed by the electron microscope. *J. Cell. Comp. Physiol.* **42,** 15.

Sjöstrand, F. S. (1953b). The ultrastructure of the inner segments of the retinal rods of the guinea pig as revealed by the electron microscope. *J. Cell Comp. Physiol.* **42,** 45.

Stephens, G. C., Fingerman, M., and Brown, F. A. (1953). The orientation of *Drosophila* to plane polarized light. *Ann. Entomol. Soc. Amer.* **46,** 75.

Strain, H. H. (1951). "Manual of Phycology." Chronica Botanica, Waltham, Massachusetts.

Strother, G. K. (1966). Absorption of *Musca domestica* screening pigment. *J. Gen. Physiol.* **49,** 1087.

Strother, G. K., and Casella, A. J. (1970). Microspectrophotometry of arthropod screening pigments. *J. Gen. Physiol.* In press.

Strother, G. K. and Wolken, J. J. (1960a). Microspectrophotometry of *Euglena* chloroplast and eyespot. *Nature (London)* **188,** 601.

Strother, G. K. and Wolken, J. J. (1960b). Microspectrophotometry. I. Absorption spectra of colored oil globules in the chicken retina. *Exp. Cell Res.* **21,** 504.

Strother, G. K. and Wolken, J. J. (1961). *In vivo* absorption spectra of *Euglena:* chloroplast and eyespot. *J. Protozool.* **8,** 261.

Tischer, J. (1936). Über das Euglenarhodon und andere Carotinoide einer roten Euglene. (Carotinoide der Susswasseralgen, I. Teil.). *Hoppe-Seyler's Z. Physiol. Chem.* **239,** 257.

Tollin, G. and Robinson, M. I. (1969). Phototaxis in *Euglena* V. Photosuppression of phototactic activity by blue light. *Photochem. Photobiol.* **9,** 411.

Vaissière, R. (1961a). Morphologie et histologie comparées des yeux des Crustacés Copépodes. Ph.D. Thèse. Centre National de la Recherche Scientifique, Paris, France.

Vaissière, R. (1961b). Morphologie et histologie comparées des yeux des Crustacés Copépodes. *Arch. Zool. Exp. Gen.* **100,** 126.

van Dorp, D. A. and Arens, J. F. (1947). Synthesis of vitamin A aldehyde. *Nature (London)* **160,** 189.

Varela, F. G. and Wiitanen, W. (1970). The optics of the compound eye of the honeybee *(Apis mellifera). J. Gen. Physiol.* **55,** 336.

von Frisch, K. (1914). Der Farbensinn und Formensinn der Bienen. *Zool. Jahrb. Physiol.* **35,** 1.

von Frisch, K. (1949). Die Polarisation des Himmelslichtes als Faktor der Orientieren bei den Tänzen der Bienen. *Experientia* **5,** 397.

von Frisch, K. (1950). "Bees: Their Vision, Chemical Senses and Language." Cornell Univ. Press, Ithaca, New York.

von Frisch, K. (1953). "The Dancing Bees." Harcourt, Brace & Co., New York.

von Frisch, K. (1967). "The Dance Language and Orientation of Bees." pp. 385–392. Harvard Univ. Press, Belknap Press, Cambridge, Massachusetts.

von Helmholtz, H. (1852). Über die Theorie der zusammemgesetzien Farben. *Ann. Phys.* **87,** 45.

von Helmholtz, H. (1867). "Hanbuch der Physiologischen Optik," p. 874. Leopold Voss, Leipzig.

Vowles, D. M. (1954). The orientation of ants. II. Orientation to light, gravity and polarized light. *J. Exp. Biol.* **31,** 356.

Waddington, C. H. and Perry, M. M. (1960). The ultrastructure of the developing eye of *Drosophila. Proc. Roy. Soc. Ser.* **B153,** 155.

Wald, G. (1938). On rhodopsin in solution. *J. Gen. Physiol.* **21,** 795.

Wald, G. (1939). On the distribution of vitamins A_1 and A_2. *J. Gen. Physiol.* **22,** 391.

Wald, G. (1945). Human vision and the spectrum. *Science* **101,** 653.

Wald, G. (1948). Galloxanthin, a carotenoid from the chicken retina. *J. Gen. Physiol.* **31,** 377.

Wald, G. (1953). The biochemistry of vision. *Annu. Rev. Biochem.* **22,** 497.

Wald, G. (1954). On the mechanism of the visual threshold and visual adaptation. *Science* **119,** 887.

Wald, G. (1955). Photoreceptor process in vision. *Amer. J. Ophthalmol.* **40,** 18.

Wald, G. (1956). The biochemistry of visual excitation. *In* "Enzymes: Units of Biological Structure and Function" (O. H. Gaebler, ed.), p. 355. Academic Press, New York.

Wald, G. (1957). The origin of optical activity. *Ann. N. Y. Acad. Sci.* **69,** 352.

Wald, G. (1959). The photoreceptor process in vision. *In* "Handbook of Physiology. I. Neurophysiology," p. 671. American Physiological Society, Washington, D. C.

Wald, G. (1961). General discussion of retinal structure in relation to the visual process. *In* "The Structure of the Eye" (G. K. Smelser, ed.), pp. 101–115. Academic Press, New York.

Wald, G. (1964). The receptors of human color vision. *Science* **145,** 1007.

Wald, G. (1965). Introductory lecture. *In* "Recent Progress in Photobiology" (E. J. Bowen, ed.), p. 333. Blackwell, Oxford/Academic Press, New York.

Wald, G. (1968). Single and multiple visual systems in arthropods. *J. Gen. Physiol.* **51,** 125.

Wald, G. and Allen, G. (1946). Fractionation of the eye pigments of *Drosophila melanogaster. J. Gen. Physiol.* **30,** 41.

Wald, G. and Brown, P. K. (1957). The Vitamin A of a euphausiid crustacean. *J. Gen. Physiol.* **40,** 627.

Wald, G. and Brown, P. K. (1958). Human rhodopsin. *Science* **127,** 222.

Wald, G. and Brown, P. K. (1965). Human color vision. *Cold Spring Harbor Symp. Quant. Biol.* **30,** 345.

Wald, G. and Burg, S. P. (1957). The vitamin A of the lobster. *J. Gen. Physiol.* **40,** 609.

Wald, G. and Hubbard, R. (1957). Visual pigment of a decapod crustacean: the lobster. *Nature (London)* **180,** 278.

Wald, G. and Seldin, E. B. (1968). Spectral sensitivity of the common prawn, *Palaemonetes vulgaris. J. Gen. Physiol.* **51,** 694.

Wald, G., Brown, P. K., and Gibbons, I. R. (1963). The problem of visual excitation. *J. Opt. Soc. Amer.* **53,** 20.

Walls, G. L. (1942). "The Vertebrate Eye," pp. 29 and 265. Cranbrook Institute of Science, Bloomfield Hills, Michigan.

Walther, J. B. (1958). Changes induced in spectral sensitivity and form of retinal action potential of the cockroach eye by selective adaptation. *J. Insect Physiol.* **2,** 142.

Walther, J. B. and Dodt, E. (1957). Electrophysiologische Untersuchungen über die Ultraviolettempfindlichkeit von Insektenaugen. *Experientia* **13,** 333.

Walther, J. B. and Dodt, E. (1959). Die Spektralsensitivität von Insekten Komplexaugen in Ultraviolette. *Z. Naturforsch.* **148b,** 273.

Waterman, T. H. (1966). Systems analysis and the visual orientation of animals. *Amer. Sci.* **54,** 15.

Waterman, T. H., Fernández, H. R., and Goldsmith, T. H. (1969). Dichroism of photosensitive pigment in rhabdoms of the crayfish *Orconectes. J. Gen. Physiol.* **54,** 415.

Weibul, C. (1951). Some analytical evidence for the purity of Proteus flagella protein. *Acta Chem. Scand.* **5,** 529.

Wigglesworth, V. B. (1949). Insect biochemistry. *Annu. Rev. Biochem.* **18,** 595.

Wigglesworth, V. B. (1964). "The Life of Insects." Weidenfeld and Nicholson, London.

Willmer, E. N. (1955). The physiology of vision. *Annu. Rev. Physiol.* **17,** 339.
Wolken, J. J. (1956a). A molecular morphology of *Euglena gracilis v. bacillaris. J. Protozool.* **3,** 211.
Wolken, J. J. (1956b). Photoreceptor structures. I. Pigment monolayers and molecular weight. *J. Cell. Comp. Physiol.* **48,** 349.
Wolken, J. J. (1958a). Studies of photoreceptor structures. *Ann. N. Y. Acad. Sci.* **74,** 164.
Wolken, J. J. (1958b). Retinal structure. Mollusc cephalopods: *Octopus, Sepia. J. Biophys. Biochem. Cytol.* **4,** 835.
Wolken, J. J. (1961a). A structural model for a retinal rod. *In* "The Structure of the Eye" (G. K. Smelser, ed.), pp. 173–192. Academic Press, New York.
Wolken, J. J. (1961b). The photoreceptor structure. *In* "International Review of Cytology" (G. H. Bourne and J. F. Danielli, eds.), Vol. 11, pp. 195–218. Academic Press, New York.
Wolken, J. J. (1963). The structure and molecular organization of retinal photoreceptors. *J. Opt. Soc. Amer.* **53,** 1.
Wolken, J. J. (1966). "Vision: Biophysics and Biochemistry of the Retinal Photoreceptors," p. 193. Thomas, Springfield, Illinois.
Wolken, J. J. (1967). "*Euglena:* An Experimental Organism for Biochemical and Biophysical Studies," 2nd ed. Appleton-Century-Crofts, New York.
Wolken, J. J. (1969). Microspectrophotometry and the photoreceptor of *Phycomyces* I. *J. Cell Biol.* **43,** 354.
Wolken, J. J. (1970). Cell and photoreceptor membranes. *In* "Physical Principles of Biological Membranes" (F. Snell, J. J. Wolken, G. J. Iverson, and J. Lam, eds.), pp. 365–382. Gordon and Breach, New York.
Wolken, J. J. and Florida, R. G. (1969). The eye structure and optical system of the crustacean copepod, *Copilia. J. Cell Biol.* **40,** 279.
Wolken, J. J. and Gallik, G. J. (1965). The compound eye of a crustacean: *Leptodora kindtii. J. Cell Biol.* **26,** 968.
Wolken, J. J. and Gross, J. A. (1963). Development and characteristics of the *Euglena* C-type cytochrome. *J. Protozool.* **10,** 189.
Wolken, J. J. and Gupta, P. D. (1961). Photoreceptor structures. The retinal cells of the cockroach eye. IV. *Periplaneta americana* and *Blaberus giganteus. J. Biophys. Biochem. Cytol.* **9,** 720.
Wolken, J. J. and Palade, G. E. (1953). An electron microscope study of two flagellates. Chloroplast structure and variation. *Ann. N. Y. Acad. Sci.* **56,** 873.
Wolken, J. J. and Scheer, I. J. (1963). An eye pigment of the cockroach. *Exp. Eye Res.* **2,** 182.
Wolken, J. J. and Shin, E. (1958). Photomotion in *Euglena gracilis.* I. Photokinesis. II. Phototaxis. *J. Protozool.* **5,** 39.
Wolken, J. J. and Strother, G. K. (1963). Microspectrophotometry. *Appl. Opt.* **2,** 899.
Wolken, J. J., Capenos, J., and Turano, A. (1957a). Photoreceptor structures. III. *Drosophila melanogaster. J. Biophys. Biochem. Cytol.* **3,** 441.
Wolken, J. J., Mellon, A. D., and Contis, G. (1957b). Photoreceptor Structures. II. *Drosophila melanogaster. J. Exp. Zool.* **134,** 383.
Wolken, J. J., Bowness, J. M., and Scheer, I. J. (1960). The visual complex of the insect: Retinene in the housefly. *Biochim. Biophys. Acta* **43,** 531.
Wolken, J. J., Forsberg, R., Gallik, G. J., and Florida, R. G. (1968). Rapid recording microspectrophotometer. *Rev. Sci. Instrum.* **39,** 1734.
Wulff, V. S. (1956). Physiology of the compound eye. *Physiol. Rev.* **36,** 145.

Wurtman, R. J., Axelrod, J., and Kelly, D. E. (1968). "The Pineal." Academic Press, New York.

Yoshida, M., Ohtsuki, H., and Suguri, S. (1967). Ommochrome from anthomedusan ocelli and its photoreduction. *Photochem. Photobiol.* **6,** 875.

Yoshizawa, T. and Wald, G. (1963). Pre-lumirhodopsin and the bleaching of visual pigments. *Nature (London)* **197,** 1279.

Young, J. Z. (1960). "Doubt and Certainty in Science." Oxford Univ. Press, Oxford.

Young, J. Z. (1962). The optic lobes of *Octopus vulgaris*. *Phil. Trans. Roy. Soc. London* **B245,** 19; The retina of cephalopods and its degeneration after optic nerve section. *Ibid.* **B245,** 1.

Young, J. Z. (1964). "A Model of the Brain." Oxford Univ. Press, Oxford.

Young, T. (1802). On the theory of light and colors. *Phil. Trans. Roy. Soc. London* **92,** 12.

Young, T. (1807). "Lectures on Natural Philosophy," pp. 315 and 613. Vol. I. W. Savage, London.

Zechmeister, L. (1944). *Cis-trans* isomerization and stereochemistry of carotenoids and diphenylpolyenes. *Chem. Rev.* **34,** 267.

Zechmeister, L. (1962). "Carotenoids: *Cis-Trans* Isometric Carotenoids, Vitamins A, and Arylpolyenes." Academic Press, New York.

Ziegler, I. (1964). Uber natürlich vorkommende Tetrahydropterine. *In* "Pteridine Chemistry" (W. Pfleiderer and E. C. Taylor, eds.), pp. 295–305. Pergamon, Oxford.

Ziegler, I. (1965). Pterine als Wirkstoffe und Pigmente. *In* "Ergebnisse der Physiologie Biologischen Chemie und Experimentellen Pharmakologie" (K. Kramer, O. Krayer, E. Lehnartz, A. v. Muralt, and H. H. Weber, eds.), Vol. 56, p. 1. Springer, New York.

Ziegler-Günder, I. (1956). Pterine: Pigmente und Wirkstoffe im Tierreich. *Biol. Rev.* **31,** 313.

Zonana, H. V. (1961). Fine structure of the squid retina. *Bull. Johns Hopkins Hosp.* **109,** 185.

SUPPLEMENTAL READINGS

Beament, J. W. L., ed. "Biological Receptor Mechanisms" Symp. Soc. Exp. Biol. No. 16. Academic Press, New York, 1962.

Bernhard, C. G., ed. "The Functional Organization of the Compound Eye." Pergamon Press, Oxford, 1966.

Bullock, T. H. and Horridge, G. A., eds. "Structure and Function of the Nervous System of Invertebrate Animals," Vols. I and II. Freeman, San Francisco, California, 1965.

Carthy, J. D. and Newell, G. E., eds. "Invertebrate Receptors." Academic Press, New York, 1968.

Cold Spring Harbor Symposia on Quantitative Biology, Vol. XXX: Sensory Receptors. Cold Spring Harbor Laboratory of Quantitative Biology, Cold Spring Harbor, Long Island, New York, 1965.

Dartnall, H. J. A. "The Visual Pigments." Wiley, New York, 1957.

Dartnall, H. J. A., ed. "Handbook of Sensory Physiology. Part I: Photochemistry of Vision," Vol. VII. Springer, New York, 1970.

Dethier, V. G. "The Physiology of Insect Senses." Wiley, New York, 1963.

Duke-Elder, S. "System of Ophthalmology: The Eye in Evolution," Vol. I. Mosby, St. Louis, Missouri, 1958.

Hyman, L. H. "The Invertebrates," Vol. I: Protozoa Through Ctenophora, 1940. Vol. II: Platyhelminthes and Rhychocoela, 1951. Vol. III: Acanthocephala and Aschelminthes

and Entroprocta, 1951. Vol. IV: Echinodermata, 1955. Vol. V: Smaller Coelomata Groups, 1955. Vol. VI: Mollusca (part 1), 1967. McGraw-Hill, New York.
Lord Rothchild, "A Classification of Living Animals." Wiley, New York, 1961.
McElroy, W. D. and Glass, B., eds. "Light and Life." Johns Hopkins Press, Baltimore, Maryland, 1961.
Mazohkin-Porshnyakov, C. A. "Insect Vision." Plenum Press, New York, 1969.
Roeder, K. D. "Nerve Cells and Insect Behavior." Harvard Univ. Press, Cambridge, Massachusetts, 1963.
Rubin, M. L. and Walls, G. L. "Fundamentals of Visual Science." Thomas, Springfield, Illinois, 1969.
Seliger, H. and McElroy, W. D. "Light: Physical and Biological Action." Academic Press, New York, 1965.
Snell, F., Wolken, J. J., Iverson, G. F., and Lam, J., eds. "Physical Principles of Biological Membranes." Gordon and Breach, New York, 1970.
Thomas, J. B. "Primary Photoprocesses in Biology." Wiley, New York, 1965.
von Frisch, K. "The Dance Language and Orientation of Bees." Belknap Press of Harvard Univ. Press, Cambridge, Massachusetts, 1967.
Waterman, T. H., ed. "The Physiology of Crustacea," Vol. II: Sense Organs, Integration and Behavior. Academic Press, New York, 1961.
Wells, M. "Lower Animals." McGraw-Hill, New York, 1968.
Wigglesworth, V. B. "The Life of Insects." Weidenfeld and Nicholson, London, 1964; also in Natural History Series, The World Publishing Co., Cleveland, Ohio.
Wolken, J. J. "Vision: Biophysics and Biochemistry of the Retinal Photoreceptors." Thomas, Springfield, Illinois, 1966.
Wolken, J. J. "Photobiology." Van Nostrand-Reinhold, New York, 1968.

APPENDIX

This brief listing is intended to show the phylogenetic position of invertebrates which have been mentioned in the text. It is in no sense a complete system, but an aid in linking the scientific name with the common name, and in what phylum or class the animal may lie. Detailed classifications are given in most zoology texts, such as "General Zoology," T. I. Storer and R. L. Usinger, McGraw-Hill, New York (1965), and in the following books listed under *Supplemental Readings* (pp. 168–169): L. H. Hyman (Vols. I–VI, 1940–1967), Lord Rothchild (1961), and M. Wells (1968).

PHYLUM Class	Order	Genus and species	Common name
PROTOZOA			
Mastigophora		*Chlamydomonas*	
		Chromulina	
		Euglena gracilis	
		Euglena heliorubescens	
		Euglena rubra	
		Euglena sanguinea	
		Haematococcus pluvialis	
		Volvox	
Sarcodina		*Amoeba*	
Ciliata		*Paramecium*	
PORIFERA			sponges
COELENTERATA			
Hydrozoa		*Hydra*	freshwater polyp
		Obelia	marine hydroid
Scyphozoa			jellyfish
Anthozoa			sea anemones
PLATYHELMINTHES			flatworms
Turbellaria		*Planaria*	
		Convoluta roscoffensis	
		Dendrocoelum lacteum	
		Dugesia lugubris	
ASCHELMINTHES			
Nematoda			roundworms

PHYLUM Class	Order	Genus and species	Common name
MOLLUSCA			
Gastropoda			snails
Pelecypoda		*Mya*	bivalves
Cephalopoda		*Sepia officinalis*	cuttlefish
		Nautilus	nautilus
		Octopus vulgaris	octopus
		Loligo pealleii	squid
ANNELIDA			segmented worms
Hirudinea			leaches
Oligochaeta		*Lumbricus terrestrius*	earthworm
ARTHROPODA			
Merostomata		*Xiphosura polyphemus*	horseshoe crab
Arachnida			spiders
		Evarcha falcata	jumping spider
		Lycosa carolinensis	wolf spider
Crustacea			
	Cladocera	*Daphnia magna*	water fleas
		Daphnia pulex	water fleas
		Leptodora kindtii	water fleas
	Copepoda	*Copilia mirabilis*	copepods
		Copilia quadrata	copepods
		Cyclops	copepods
	Decapoda	*Cambarus*	crayfish
		Procambarus	crayfish
		Orconectes virilis	crayfish
		Cardisoma guanhumi	land crab
		Libinia emarginata	spider crab
		Callinectes	swimming crab
		Homarus americanus	lobster
		Palaemonetes vulgaris	prawn
		Euphausia pacifica	shrimp
		Meganyctiphanes norvegica	shrimp
Insecta			
	Diptera	*Calliphora erythrocephala*	blow fly
		Drosophila melanogaster	fruit fly
		Musca domestica	house fly
		Blaberus giganteus	cockroach
		Blatella germanica	cockroach
		Blatta orientalis	cockroach
		Periplaneta americana	cockroach
	Orthoptera	*Chorthippus*	grasshopper
		Melanoplus	locust
	Coleoptera	*Photuris pennsylvanica*	firefly
	Hymenoptera	*Vespa maculata*	bald face hornet
		Camponatus herculenus pennsylvanicus	carpenter ant

PHYLUM Class	Order	Genus and species	Common name
		Solenopsis saevissima	fire ant
		Apis mellifera	honeybee
		Tapinoma sessile	house ant
	Odonata	*Aeschna*	dragonfly
	Lepidoptera	*Tineola bisiella*	clothes moth
		Epargyreus clarus	skipper butterfly
ECHINODERMATA			sea urchins, starfish

SUBJECT INDEX

A

Acetylcholine, 42
Adenosine triphosphate (ATP), 42
Algae, 4, 6, 7, 21, 42
Amino acids, *see* Protein
Amphibia, *see* Amphibian
Amphibian, 4, 92, 98, 99, 127, 144, 148, *see also* Frog, Mudpuppy
Annelids, 49, *see also* Earthworm
Ant, 50, 65, 151, 152, *see also* Carpenter ant, Insects
Antheraxanthin, *see* Carotenoids, Xanthophyll
Antimony trichloride, 102, 127, 131
Apis mellifera, 64, 119, 122, *see also* Honeybee
Apposition eye, 51, 64, 65, 69, 77, 141
Arachnid, 50, 141, 148, *see also* Spider
Arthropod, 4, 47, 50, 69, 77, 78, 81, 84, 86–88, 90–92, 99, 111, 115, 134, 139–141, 144, 148, 149, 152, *see also* Insects, Crustacea, Arachnids
Astacene, *see* Carotenoids, Xanthophyll
Astaxanthin, *see* Carotenoids, Xanthophyll
Axon, 58, 65
Axonemata, 23

B

Bacteria, 1, 23
  photosynthetic, 4, 6, 8, 20
Bacteriochlorophyll, 6, 8, 20, *see also* Chlorophylls
Bald face hornet, *see* Hornet
Bee, 119, 120, 122, *see also* Honeybee
Bipolar cell, 93
Birds, 134
Birefringence, 94, *see also* Polarized light
Blaberus giganteus, 54, 129, *see also* Cockroach
Blatella germanica, 55, *see also* Cockroach
Blatta orientalis, 55, 129, *see also* Cockroach
Blowfly, 137–139, 150, 153, *see also* *Calliphora*
Bovine rod, 96, 97, *see also* Retinal rods, Cattle
Brain, 75, 80, 86, 87, 107
Butterfly, *see* Skipper butterfly

C

Callinectes, 74, *see also* Crab
Calliphora erythrocephala, 137–139, 151, 153, *see also* Blowfly
Camponatus herculenus pennsylvanicus, 64, *see also* Carpenter ant
Carr-Price Reagent, 102, *see also* Antimony trichloride
Cardisoma guanhumi, 74, *see also* Crab
Carp, 107
Carpenter ant, 64, 65, *see also* Ant
Carotenoids, 5, 8–12, 15, 17, 37, 38, 41, 93, 119, 127, 134, 141, 148, 155
  α-carotene, 10
  β-carotene, 8, 9, 11, 19, 36–38, 41, 102, 141
  γ-carotene, 38
  xanthophyll, 8, 10, 36, 93, 127
    antheraxanthin, 38
    astacene, 10, 37, 38
    astaxanthin, 10, 36–38
    lutein, 8, 10, 37, 38, 127

Carotenoids (*continued*)
neoxanthin, 37, 38
zeaxanthin, 38
Catecholamine, 118
Cattle, 95, 109, 111, 122, 124
rhodopsin absorption spectrum, 103
Cephalopod, 48, 49, 86, 88, 112, 114, 115, 137, 143, 148, 149, *see also* Mollusc
Chalmydomonas, 20, 22, 28
Chloroform, 102 ,127
Chlorophyll(s), 5, 7–9, 14, 15, 20, 23
Chlorophyll a, 2, 5–8, 15, 16
Chlorophyll b, 2, 5–7, 16
Chlorophyll c, 6
Chlorophyll d, 6
Chlorophyll e, 6
Chloroplast, 4, 20, 21, 23, 38, 144
Choroid, 48, 93
Chromatium, 14, *see also* Bacteria
Chromatography, 19, 38, 102, 122, 125, 127, 130, 131
Chromatophores, 4, 20, 87
Chromophore, 2 111–113, 126, 137, 140, 148, *see also* Prosthetic group
Chromulina, 45, *see also* Chrysomonad, Flagellates
Chrysomonads, 6, 45
Cilia, 23, 97, 141, 154
Clam, 4, 86
Clothes moth, 62, *see also* Moth
Cockroach, 54, 55, 57, 58, 89, 91, 129, 131, 132, 143
Coelenterates, 4, 47
Color blind, 117
Color vision, 106, 107, 116, 118, 133, 139, 149, 150
Compound eyes, 4, 47, 48, 50–52, 54, 57, 58, 61, 64, 65, 68–70, 74, 75, 78, 87, 88, 112, 131, 139, 141, 143, 152–154
fast-type, 90, 91, 143, 144
slow-type, 90, 91, 143, 144
Cone pigment, 106, 107, 111, 116, 150, *see also* Iodopsin, Cyanopsin
Convoluta roscoffensis, 50, *see also* Flatworm, Planaria
Copepod, 74, 77
Copilia, 77, 78, 81, 83, 84, 86, 87, 89, 143, *see also* Copepod, Crustacea
Copilia mirabilis, Caribbean, 77, 83, 85
Copilia quadrata, Mediterranean, 77, 80, 83, 85
Cornea, 48, 87
Corneal lens, 49–52, 58, 69, 71, 78, 81, 83, 85, 132, 136, 139, 141, 152, *see also* Lens
Crab, 70, 74
Crayfish, 116, 138, 150, 153
Crustacea, 4, 19, 36, 37, 50, 69, 70, 74, 77, 84, 89, 91, 115, 117, 137, 141, 143, 148–150
Crystal, 102, 144, 155
Crystalline cone, 50, 51, 58, 62, 69, 71, 75, 77, 78, 80–82, 117, 136, 141, 143
Crystalline cone thread, 58, 59, 65, 68
Crytomonads, 6
Crytoxanthin, 38
Cuttlefish, 87, 114, 143, *see also Sepia*
Cyanopsin, 100, 101, 106, 150, *see also* Cone pigment
Cyclops, 74
Cysteine, 42
Cytochrome, 7, 12, 14–16, 135
Cytopharynx, 23

D

Daphnia, 70, 74, 84, 147, *see also* Water flea
Daphnia magna, 70
Daphnia pulex, 70
Dendrocoelum lacteum, 49, *see also* Planaria
Dermal light sense, 4
Diatom, 6
Dichroism, 94, 153, *see also* Polarized light
in rhabdom, 152–154
Digitonin, 103, 112, 119, 125, 126, 130, 149
Dihydroxyphenylalanine (DOPA), 15
Dragonfly, 50, 77, 90, 143, 144
Drosophila, 57, 59, 89, 91, 117–119, 143, *see also* Fruitfly
Drosophila melanogaster, 52, 83, 117
Dugesia lugubris, 49, *see also* Planaria

E

Earthworm, 4, 49
Echinenone, 38

Electron microscopy, 17, 19, 23, 41, 45, 52, 54, 58, 87, 89, 92–95, 154
Electroretinogram (ERG), 68, 90, 115, 117
Epargyreus clarus, 143, *see also* Skipper butterfly
Epidermis, 154
Euglena, 6, 20, 21, 23, 27, 28, 31, 33, 34, 37, 38, 41, 42, 44–46, 97, 140, 141
Euglena gracilis, 23, 32, 37
Euglena heliorubescens, 36
Euglena rubra, 36
Euglena sanguinea, 36
Euglenanone, 38
Euglenarhodon, 36
Euglenoids, *see Euglena*
Euphausia pacifica, 115, *see also* Shrimp
Euphausiid, 115, 116, *see also* Shrimp
Evarcha falcata, 133, *see also* Jumping spider, Spider, Arachnid
Exocone, 58, *see also* Crystalline cone
Exoskeleton, 23, 74, *see* Pellicle
Eye, *see* Refracting eye, Compound eye, Pine-hole eye
  structure, 142
Eyespot, 4, 20, 23, 27, 31, 33, 38, 39, 41–47, 50, 112, 140, 141, 154
Eyespot–flagellum, 28, 44–47, 141, 154
Eyespot pigment, 36–48
  absorption spectra, 40

F

Fern, 6
Ferredoxin, 14–16
Fibril, 23, 27, 41–43, 45, 97, *see also* Flagellum, Cilia
Firefly, 57–59, 64, 68, 89, 143
Fish, 127, 148
Flagellar protein, 42
Flagellates, 6, 20, 21, 141
Flagellum, 20, 23, 27, 33, 38, 39, 41–43, 45, 46, 96, 97, 140, 141, 154
Flatworm, 4, 47, 49, *see also* Planaria
Flavine semiquinone, 38
Flavines, 14, 15, 17, 41, 151, *see also* Riboflavin
Flavoprotein, 14, 39, 41, 141, 148
Flowering, 1, 3, 11
Fovea, 48, 93, 107, *see also* Human eye
Frog, 95, 96, 98, 99, 102, 103, 107–111, 153, *see also* Amphibia
  rhodopsin absorption spectrum, 104
Fruitfly, 52, 119, *see also Drosophila*
Fungi, 21, 147, *see also Phycomyces*

G

Ganglion cell, 4, 88, 93, 118
Gold fish, 107
Grasshopper, 69, 143
Guinea pig, 94
Gullet, 23, 27, 39, 140

H

Haematochrome, 36
Haematococcus pluvialis, 36, *see also* Flagellates
Hering's theory, 107, *see also* Color vision
Homarus americanus, 115, *see also* Lobster
Honeybee, 64, 92, 111, 119, 120, 122, 129, 137–139, 143, 149, 151, 152
  drone, 69, 139, 151
  worker, 120, 151
Hornet, 61, 63–65, 89, 143
Horseshoe crab, 133, *see also* King crab, *Limulus*
Housefly, 54, 122–127, 129, 136–138, 149, *see also Musca domestica*
Human eye, 32, 93, 107, 150
  retina, 93
  spectral sensitivity, 100
  structure, 48

I

Indicator yellow, 113, 114, *see also* Metarhodopsin
Insects, 2, 19, 46, 50–52, 54, 57, 58, 61, 65, 68, 70, 71, 83, 89–91, 117–120, 122, 123, 125, 129, 137, 139, 141, 143, 144, 148–152
Invertebrate visual pigments, 113–133, 138
Iodopsin, 100, 101, 106, 116, 150, *see also* Cone pigment
Isoalloxazine, 14, *see also* Riboflavine
Isoprene, 8, 10

J

Jumping spider, 132, *see also Evarcha falcata*, Spider

K

Ketocarotenoids, *see* Carotenoids
  euglenanone, 38
  hydroxyechinenone, 38
King crab, 132, 133, *see also* Horseshoe crab, *Limulus*
Kynurenine, 118

L

Leech, 49
Lens, 34, 48, 49, 51, 75, 78, 81, 83, 85, 87, 132, *see also* Corneal lens
Leptodora 74–77, 89, 117, *see also* Water flea
Leptodora kindtii, 74, 84
Libinia emarginata, 116, *see also* Spider crab
Limulus polyphemus, 133, *see also* Horseshoe crab, King crab
Lipids, 94, 109–111, 146, 155
Lipoprotein, 109–112, 149
Lizard, 4, 134
Lobster, 36, 70, 115, 138, 150
Locust, 69, 143
Loligo pealeii, 112, *see also* Squid
Lumichrome, 14, *see also* Riboflavine
Lumirhodopsin, 105, 106, *see also* Rhodopsin
Lutein, *see* Carotenoids, Xanthophyll
Lycosa, 151, *see also* Wolf spider

M

Macula lutea, 93, *see also* Fovea
Magnesium, 5, 7
Mastigonemata, 23
Median eye, 133
Meganyctiphanes norvegica, 116, *see also* Shrimp
Melanin, 15, 17, 118, 125, 151
Melanophore, 4
Membrane, 23, 27, 51, 52, 54, 55, 58, 71, 88, 96, 109–112, 144, 149, 153–155
Metarhodopsin, 105, 106, 112, 113, 115, 123, 124, 138, 149, *see also* Rhodopsin
Metazoa, 21
Microscopy, *see also* Electron microscopy
  interference, 17
  phase, 17
  polarizing, 17, 93, 94
Microspectrophotometer, 19, 38, 129, 151, *see also* Spectroscopy
Microspectrophotometry, 19, 107, 108, 112, 116, 117, 125, 136, 137, 139, 141, 149, 153, *see also* Spectroscopy
Microtubules, 54, 57, 64, 65, 71, 74, 77, 81, 87–90, 92, 111, 144, 149, 153
Microvilli, 144
Mitochondria, 55, 80, 96, 97
Molluscs, 4, 19, 47–49, 69, 70, 86, 89, 91, 92, 99, 112, 115, 140, 141, 143, 144, 148, 149, 152, *see also* Cephalopod
Molybdenum, 118
Monkey, 95, 107
Moss, 6
Moth, 58, 62, 143
Mudpuppy, 99, *see also Necturis*, Amphibia
Musca domestica, 54, 83, 122, 136, *see also* Housefly
Muscle, 42, 45, 81
Mya, 4, *see also* Clam
Myosin, 42

N

Nauplius eye, 70
Nautilis, 48, 86, *see also* Cephalopod, Mollusc
Necturis, 99, *see also* Mudpuppy, Amphibia
Neoxanthin, *see* Carotenoids, Xanthophyll
Nerve bundle, 65
Nerve cell, 4, 42, 62, 93
Nerve excitation, 42, 107
Nerve pulse, 45, 46, 49, *see also* Synapse
Nerve vessicle, 80
Nervous system, 52
Neurofibrillar network, 49
Neuron, 65, 74
Nucleic acids, 2

O

Ocelli, 4, 49, 50, 112, 141, 154, *see also* Simple eye
Ocellus, *see* Ocelli
Octopus, 49, 86–88, 114, 115, 143
Octopus vulgaris, see Octopus
Oil globules, 134, 139
Ommatidium, 50–52, 54, 57–59, 61, 64, 65, 68, 69, 71, 75, 77, 78, 81, 88, 136, 137, 141
Ommatines 134
Ommines, 134
Ommochrome, 119, 125, 134, 139
Opic ganglion, 118
Opsin, 100, 103–106, 110, 112–116, 130, 138, 148, 149
Optic nerve, 46, 48, 80, 81
Orconectes virilis, 116, 153, *see also* Crayfish
Oyster, 86
Ozone, 2

P

Palaemonetes vulgaris, 116, *see also* Prawn
Paraflagellar structure, 27, 38, 141
Pellicle, 23, *see also* Exoskelton
Perch, 95
Periplaneta americana, 54, 55, 129, 132, *see also* Cockroach
Petroleum ether, 5, 38, 102, 127, 131
Pheophytin, 7, *see also* Chlorophyll
Photocell, 44, 45, 140
Photoconductivity, 146
Photokinesis, 27, 28, 32, 34, 35, 44
Photomotion, 27, *see also* Photokinesis, Phototaxis
Photons, 3, 31, 32
Photoperiodism, 3, 11
Photoreceptor(s), 1, 4, 5, 17, 19, 20, 27, 28, 31, 37, 41, 44–48, 50, 57, 58, 61, 65, 68, 69, 78, 87, 91, 93, 100, 111, 112, 129, 133, 136 137, 139–141, 143, 144, 146, 148, 150, 151, 154
Photoreceptor evolution, 154
Photoreceptor structure, 47, 50, 69, 89, 92, 117, 141
Photosensory cell, 4, 44, 47, 49
Photosynthesis, 1, 3, 4, 14, 16, 20, 21, 23, 45
Phototaxis, 1, 27, 28, 33–36, 39, 41, 44, 45
Phototropism, 1, 2, 45, 148
Photuris pennsylvanica, 57, 58 ,143, *see also* Firefly
Phycomyces blakesleeanus, 147, *see also* Fungi
Phytochrome, 11–13
Phytol, 5, 6, 8–10, *see also* Chlorophyll
Pigment cell, 49, 51, 58, 62, 77, 87
Pigment granule, 45, 49, 51, 54, 55, 71, 80, 133, 134, 136, 137
Pineal organ, 4
Pin-hole eye, 47, 48, 86, 141
Planaria, 49, 50, *see also* Flatworm
Plastids, 4, *see also* Chloroplast
Platyhelminthes, 49, *see also* Flatworm, Planaria
Polarized light, 28, 30, 35, 74, 91, 151–153
  analyzer in eye, 92, 152
  von Frisch's model, 92
Porphyrin, 5, 7, 8, 12
Porphyropsin, 100, 101, *see also* Retinal rod
Prawn, 116
Procambarus, 116, *see also* Crayfish
Prosthetic group, 12, *see also* Chromophore
Protein, 2, 12, 14, 42, 94, 103, 106, 109–112, 115, 124, 137, 140, 146, 149, 151, 155
Protochlorophyll, 7, 8, *see also* Chlorophylls
Protozoa, 4, 19–21, 47, 91, 140, 141, 148, 154
Pseudocone, 58, 64, *see also* Crystalline cone
Pteridine, 15, 118, 119, 134, 139, 151
Pterine, 134, 135
Purkinje shift, 118

Q

Quanta, 3, 46
Quinone, 15, 134

R

Rabbit, 95
Radiation
  electromagnetic, 1, 3

Radiation (*continued*)
infrared, 3, 19
ionizing, 1, 23
radio waves, 1
solar, 2
ultraviolet, 2, 14, 19, 23, 38, 113, 117, 119, 120, 124, 132, 133, 135, 137–139, 150, 151
Rana pipiens, 97, *see also* Frog, Amphibian
Rat, 95, 111
Refracting eye, 47–49, 86, 112, 141, 154, *see also* Human eye, Vertebrate eye
Refractive index, 81, 85, 94, *see also* Lens
Retina, 1, 4, 36, 37, 48, 87, 88, 92, 93, 98, 100, 102, 106, 113, 119, 139, 143, 153
Retinal, 19, 100, 101, 103–106, 109–115, 119, 120, 122, 126, 127, 129–131, 138, 148, 149, 151
isomers, 101
$Retinal_1$, 11, 100–102, 106, 115, 116, 120, 122, 127, 131, 133, 137, 148–150
$Retinal_2$, 100–102, 106, 148, 150
Retinal cell, 23, 46, 93, 144
Retinal cone, 4, 45, 87, 93, 96, 100, 107, 118, 134, 139, 144, 149, 150
Retinal photoreceptor, 19, 45, 49, 87, 93
Retinal rod, 4, 5, 45, 49, 50, 87, 88, 92–100, 103, 107–112, 139, 141, 144, 149, 150, 153, *see also* Vertebrate rod
Retinene, *see* Retinal
Retinol, *see* Vitamin A
Retinula cell, 49–52, 54, 55, 58, 61, 64, 65, 68, 69, 71, 77, 78, 80, 87, 89, 117, 138, 141, 143, 144, 151, 153, *see also* Ommatidium
Retinylideneamine, 123, *see also* Indicator yellow
Retonic acid, *see* Vitamin A acid
Rhabdom, 49–51, 54, 55, 57, 58, 61, 64, 65, 71, 75, 77, 78, 80, 81, 83–85, 87–92, 99, 116, 117, 125, 129, 136, 137, 139, 141, 143, 144, 151–154, *see also* Compound eyes
Rhabdomere, 50–52, 54, 55, 57, 58, 61, 64, 65, 68, 71, 74, 77, 78, 80, 81, 83, 84, 87–90, 92, 129–131, 137, 139, 141, 143, 144, 149, 151–153
Rhodommatin, 125
Rhodopsin, 2, 12, 39, 41, 46, 100, 101, 103–119, 122, 125, 129–133, 137, 141, 148–153
bleaching, 106
molecular weight, 108, 111
Rhodospirillum rubrum, 14, *see also* Bacteria
Riboflavine, 14, 15, *see also* Flavines
Roundworm, 47

S

Scanning eye, 77
Schwann cell, 58
Screening pigment, 36, 41, 69, 118, 119, 133–137, 139, 141, 150, 151
granules, 69, 81, 87, 133, 134, 137, 139, 151
Sepia, 88, 89, *see also* Cuttlefish
Sepia officinalis, 87, 114, *see also* Sepia, Cuttlefish
Shrimp, 115, 116
Simple eye, 50, 70, 154, *see also* Ocellus
Skipper butterfly, 143
Snake, 134
Spectroscopy, methods, 19, 34, 103, 154, *see also* Microspectrophotometry
Spider, 132, 133, *see also* Arachnid
Spider crab, 116, 118
Spinach, 14
Squid, 49, 86, 88, 89, 112, 113, 123, 143
Stigma 23, *see also* Eyespot
Streptomycin 23
Sulfosalicylic acid, 124
Superposition eye, 51, 57, 65, 69, 141, 143, *see also* Compound eyes
Superposition image, 77
Synapse, 62, 118, *see also* Nerve pulse

T

Tetrapyrrole, 5
Tineolo bisselliella, 64, see also Clothes moth
Tryptophan, 118
Turtle, 134
Tyrosine, 15, 118

U

Ultraviolet radiation, *see* Radiation

V

Vertebrate, 4, 19, 48, 52, 87, 88, 93, 99, 100, 103, 106, 111–116, 118, 119, 122–124, 127, 131, 133, 134, 138–141, 143, 144, 148–150, 153, 154
Vertebrate eye, 11, 50, 87, 100, 122, 138, 143, 150, 154
Vertebrate rod, 54, 92, 95, 111, 116, 131, 138, 141, 153, *see also* Retinal rod
Vespa maculata, 61, 63, *see also* Hornet
Vision, 1, 4, 45, 50, 52, 65, 68, 86, 100, 118, 143
Visual pigment 11, 12, 39, 41, 69, 92–94, 98, 100, 103, 107, 109, 111, 112, 114–120, 122–124, 126, 129, 132, 133, 137–139, 141, 144, 148–152
  absorption spectra, 100
Visual sensitivity, 92
Visual spectral sensitivity, 2
Visual threshold, 100
Vitamin A, 9, 11, 37, 100–102, 104, 105, 115, 116, 119, 122
Vitamin A_1, 9, 11, 19, 100–102, 122, 127
Vitamin A_2, 100, 101, 127
Vitamin A acid, 100
Vitamin A aldehyde, 100, 101, *see also* Retinal
Vitamin B_2, 14, *see also* Riboflavine
Volvox, 28

W

Water flea, 70, 74, 84, *see also Daphnia, Leptodora*
Waveguide, 68, 143, *see also* Crystalline cone thread
Wolf spider, 133, 151
Worm, 47

X

Xanthommatin, 15, 134, 135, 137, 150, 151
Xanthophyll, *see* Carotenoids
Xanthopterin, 135
X-ray diffraction, 110, 154

Z

Zeaxanthin, *see* Carotenoids, Xanthophyll

14 DAY